the ART of INVENTION

"We live in a time when young people seem increasingly disinclined to explore possible careers in engineering. Even those who do choose to enter the field often do not realize its tremendous creative potential. Paley does a brilliant job of communicating both the process and the joy of invention in a way that reaches both professionals and newcomers equally well. While making it clear that the inventive process cannot be reduced to a menu of tasks, he properly asserts that simplicity, elegance, and robustness are key components of the successful design process. By looking at items most of us take for granted (the paper clip, Velcro, etc.), we get to see the beauty in these designs—a simple beauty that can be found in numerous other artifacts in common use, as well as in those yet to be invented. This book goes far beyond cataloging the inventions of others, and explores the process by which you can unlock your own creativity for the solution of challenges facing us both as individuals and in the broader context of the marketplace. Before long, you will have your own sketchpad out and be coming up with your own ideas!

"The idea of the 'mad inventor' is elegantly replaced with invention as a skill within the grasp of us all. If you are willing to give yourself the freedom to tinker, to observe, to mess around with ideas, some of which may not work the first time, this book should be in your hands—not resting on a shelf somewhere."

—David Thornburg, PhD, director, Thornburg Center for Space Exploration
and former adjunct faculty member in the
Stanford University Design Division

"A must-have for any inventor. Easy to read and understand with simple, clear examples. *The Art of Invention* teaches creative thinking in a way that inspires us to invent. Through numerous practical examples and fascinating case studies, Steven Paley guides us on a journey through the world of creativity in engineering. This is the first book I have ever seen that shows both the aspiring and the experienced inventor a clear path to successfully fulfill their goals."

—Ronald D. Fellman, PhD
Former professor of electrical and computer engineering at the
University of California at San Diego, lifelong inventor, and founder of QVidium
and Path1 Network Technologies

"Steven Paley's *The Art of Invention* tells it like it is. This is an excellent introduction to the psychology of the inventor and to the nature of the inventive process."

—Henry Petroski
Author of *The Essential Engineer:*
Why Science Alone Will Not Solve Our Global Problems

the ART of INVENTION

THE CREATIVE PROCESS OF DISCOVERY AND DESIGN

STEVEN J. PALEY

59 John Glenn Drive
Amherst, New York 14228-2119

Published 2010 by Prometheus Books

Inquiries should be addressed to
Prometheus Books
59 John Glenn Drive
Amherst, New York 14228–2119
VOICE: 716–691–0133
FAX: 716–691–0137
WWW.PROMETHEUSBOOKS.COM

14 13 12 11 10 5 4 3 2 1

Library of Congress Cataloging-in-Publication Data

Paley, Steven J., 1955–.
The art of invention : the creative process of discovery and design / by Steven J. Paley.
p. cm.
Includes bibliographical references and index.
ISBN 978–161614–223–0 (pkb. : alk. paper)
1. Engineering. 2. Inventions I. Title.

T47.P25 2010
601/.9—dc22

2010029634

Printed in the United States of America

To my father, Edward Paley,
a great inventor and entrepreneur
who taught me everything about the art of invention.

"The Road Not Taken" from The Poetry of Robert Frost, *edited by Edward Connery Latham.*

CONTENTS

PART III. MAKING IT HAPPEN

INTRODUCTION

As a young boy, I used to travel with my father in search of ideas. He would take me along as he attended trade shows for everything from automobile supplies to dental equipment. My father was never looking for anything in particular but would simply walk the aisles, stopping at any booth that displayed something that aroused his curiosity. He was cataloging possibilities—assorted things that might one day come in handy. He was building up a library of ideas in his mind. My father was an inventor and an entrepreneur before the word entered the American lexicon. He was always in search of the next great idea and he attended trade shows to stimulate his thinking. He never knew what he was looking for, but sometimes he would just find it.

This is where I had my introduction to the art of invention. This is where I learned how the creative mind works. I came to realize that not everything was connected by a straight line, and that our brains work in ways that we cannot always fathom. Creativity is the process of making order out of the seemingly random. I eventually studied engineering in college and then, in graduate school, something called "Product Design," which was a combination of engineering, design, and entrepreneurship. It was during graduate school that the experiences from my youth came into the fore. The idea of "wandering around without an obvious destination" was actually considered something good. Daydreaming was even encouraged! In fact, many ideas that fly in the face of convention were given validation. Throughout my subsequent career in engineering and business, I put this mindset to work with much success.

Many of the ideas proposed in this book fly in the face of convention. They are not taught in school and are in some ways anathema to the ways we normally go about our lives. But then, inventing is different. It is creating something new, something that has never been done before. Learning to be an inventor requires unlearning. We need to unlearn some of the constraints under which we typically approach problems. We have to cast aside a certain amount of our rationality in favor of the more mysterious and ephemeral ways of our subconscious mind, and we need to learn to move back and forth between the two. The process of inventing is an exciting journey where the mind's creativity is harnessed and applied to produce a novel solution to a problem. As with many endeavors, the process is as fulfilling as the final product.

I was imbued with the passion for inventing from an early age. Walking through trade shows with my father, moving from booth to booth examining wares and products produced a wonder in me for all that was new—all the interesting things that people could create. I witnessed firsthand how these seemingly aimless wanderings resulted in the generation of new ideas, as my father synthesized all the seeing, touching, and talking into new inventions to solve problems he identified in the marketplace.

This book is about how to invent. Inventing is a creative act, and therefore it would be impossible to write a cookbook on how to invent with "recipes for invention." Rather, in the coming chapters, I examine the art of invention. How do you come up with ideas? How do you sustain and refine a vision? What qualities do most inventors have? What constitutes a good invention? Immersing yourself in the coming chapters is nothing if not a personal journey through which you can hone your own creativity and vision.

This book is divided into three sections: the first introduces the principles of invention and creative problem solving. The principles are illustrated through descriptions and case histories of many actual inventions. The second part focuses on design as the embodiment of invention. Here I elaborate on three core ideas: simplicity, elegance, and robustness. The final section of the book discusses practical matters, such as the nitty-gritty details of iterative problem solving as an idea is taken from concept to reality, as well as how to commercialize an invention.

I begin with a discussion of the paper clip. The association normally conjured when one says *paper clip* is not "great invention." Yet this simple piece of bent wire has so much to teach us about the art of invention. The paper clip embodies all the principles that make an invention great. This may not be obvious to you now, but it will be explored in great detail in the coming pages. What is it that makes the paper clip a great invention? Perhaps you will say, "But it's so simple!" or "It's almost trivial." But is it? Many of the inventions described in this book are now taken for granted because they have become so ubiquitous in our everyday lives. However, that is the mark of success.

This book is written for the aspiring inventor. I think of the high school or college student with a technical bent who is interested in creating the future. I remember how liberating it was for me to realize that it was okay not to walk the straight and narrow path—that I could open up my mind and create what I could imagine. My goal with this book is to enable you to do the same.

PART I

THE PROCESS OF INVENTION

Chapter 1

THE PAPER CLIP AND THE PROBLEM

What does it take to invent? How do you invent something—create something that didn't exist before you brought it into the world? How do people come up with great ideas?

WHAT MAKES A GREAT IDEA GREAT?

Let's start by looking at an example of a great idea—the paper clip. What makes the paper clip such a great invention? First, it solves a significant and obvious problem: temporarily holding sheets of paper together without deforming them in any way. Before the advent of computers and electronic communication, paper was, and arguably still is, the primary means of written communication. A simple way to hold together sheaves of paper was a significant need for anyone who dealt in written documents. The fact that this method—unlike stapling, pinning, or binding—was temporary provided another valuable feature in that papers could be easily reordered, added, or removed from the bound stack. While all this is nice, it still doesn't explain what is so intriguing about the simple paper clip.

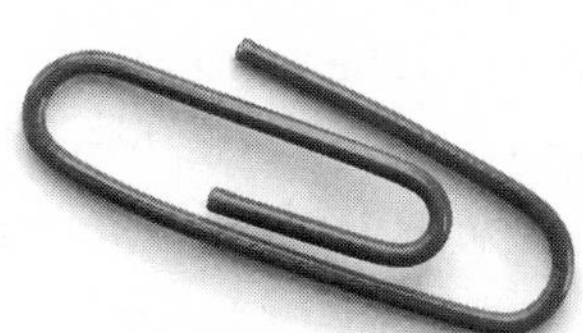

Figure 1.1: The "ordinary" paper clip. *Photo by Ivan Boden.*

Let's look at the paper clip in terms of what it can do. My first assignment in engineering school was to find fifty uses for a paper clip outside its intended use. This was not very difficult. Some of the more creative uses I have seen include a device to pick locks, a linkage to fix a broken automobile transmission, and a heated knife to sculpt foam. This simple invention provides the germ of many more inventions. That we can do so many things with something so simple begins to show why the paper clip is such a great invention.

Is it hard to make? Is it costly? The paper clip is simple in design, materials, and manufacture. It is merely a piece of extruded metal wire with three bends. And yet it can be used to do so many things.

Let's look at some of the characteristics of the paper clip as an invention.

1. **Simplicity.** The paper clip is an extremely simple solution to the problem of temporarily holding sheaves of papers together.
2. **Adaptability.** The paper clip's simplicity gives it *flexibility in use* that can be adapted to variation. Whether you want to bind two sheets of paper or twenty, the paper clip can be bent to accommodate the need.
3. **Ease of use.** The paper clip is easy and intuitive to use. No instruction manual is needed. It is obvious what to do with it. Its obviousness comes from its *multisensory* appeal. One can figure out how to use it by sight or by feel.
4. **Robustness.** The paper clip always works. Its simplicity ensures that failure will be rare. A user can determine immediately by sight whether a particular clip is likely to fail.
5. **Unintended functionality.** The fifty *other* things to do with a paper clip are an added bonus. *Simplicity* often leads to *universality*. The paper clip is merely a bent piece of wire, and there are many things that can be done with a piece of wire. Perhaps this is the genius of the design: that we can do *so much with so little* is what makes this an exceptional invention.
6. **Elegance.** Elegance is a combination of the above characteristics. Elegance means achieving a task by doing a lot with just a little. Elegance in design or invention means solving a problem in a very simple yet comprehensive manner.

The attributes above can be encapsulated in my personal design mantra. The goal of any design is to be *simple*, *elegant*, and *robust*. This applies to complex inventions and designs as well. No matter if I am designing a shoelace nib or a nuclear power plant, the *simple*, *elegant*, and *robust* guidelines apply. Addressing the seeming paradox of making complex things simple, Albert Einstein, a man who more than dabbled in complex things, is reputed to have famously said, "Everything should be made as simple as possible, but not simpler."

Even in developing complex theories of the workings of the universe, Einstein strove to be as simple as possible. As he also said, "Any intelligent fool can make things bigger, more complex, and more violent. It takes a touch of genius—and a lot of courage—to move in the opposite direction."

The ideas of simplicity, elegance, and robustness apply both to invention and to its close cousin, design. As we will see in the following chapters, inventions that fit these criteria are often considered the most profound and successful.

THE HISTORY OF THE PAPER CLIP

Who invented the paper clip, and how did he come up with this great idea? There was not one individual who had an "aha!" moment and invented the paper clip as we know it today. The original embodiment is thought to go back to Byzantine times, when a form of the paper clip was fashioned from brass to hold together very important documents. Unlike today's paper clips, these were not inexpensive, mass-produced throwaways. The wire paper clip was first patented in 1867 by Samuel B. Fay, whose intention was to create a device that would hold tickets to fabric. However, he noted in his patent application that his clip could also be used to hold papers together. The popularity of such clips

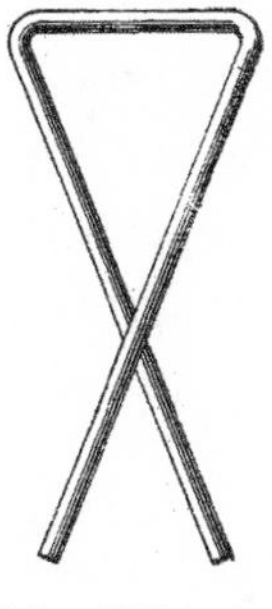

No. W174.

Figure 1.2: Samuel B. Fay's 1867 "ticket holder." *Photo courtesy of the Early Office Museum.*

for holding papers together soon overtook Fay's original interest in clipping price tags or laundry tickets to clothing. A patent for another paper clip design was issued in 1877, and patent applications for several more designs were filed in 1896 and for several years thereafter. By the 1890s, paper clips were commonly used in business offices. The March 1900 issue of *Business* commented that "[t]he wire clip for holding office papers together has entirely superseded the use of the pin in all up-to-date offices."[1]

Several interesting things emerge from looking at the question of how Fay came up with this invention. First, the original invention was designed to solve a much narrower problem than the one it actually solved. The inventor did not recognize the potential of his invention. This is commonly the case. An inventor looks to solve one problem and inadvertently solves a much larger one. There are many examples of this in the history of invention. The inventor of chewing gum, for instance, was originally trying to develop synthetic rubber from chicle sap. Leo Baekeland, the inventor of Bakelite, the first synthetic plastic, was trying to develop an alternative to shellac for insulating electrical wires.

Second, we can see that inventions evolve. Inventors are still searching for means to improve the paper clip. They identify problems with what seems to be a mature design and try to find novel improvements. The initial idea is repeatedly modified and refined until the change reaches a point of diminishing improvement. Even after that point was supposedly reached and paper clip design seemed to have reached an ideal, multicolored plastic-coated paper clips were introduced as a new variation on a theme. This is an example of looking at an invention from a different perspective and seeing how it can be refined.

Finally, and very importantly, invention depends on technology. The paper clip could not have been invented and popularized fifty years earlier. Its design and mass production required inexpensive steel wire and machines that could cheaply, rapidly, and efficiently cut and bend it into useful shapes.[2]

Without the technology to inexpensively mass-produce paper clips, they would have remained a novelty item. Technology not only makes invention possible, it makes it relevant. In our time, the technology of the personal computer created an entire industry of application software. Had computers never progressed beyond large expensive mainframes, this industry would not have been born.

What's the Problem?

The more specific and well defined the problem, the clearer the solution. If my problem statement is "I want to create an end to war," it will be very difficult to generate a clear and realistic solution. If my problem is "I want to develop a means of attaching a ticket to fabric," then solutions are easier to visualize. *Constraints help to produce creative solutions.* Boundaries provide clarity to the thought process. To invent, we need to think about the very specific problem we are solving. Paradoxically, the more sharply our problem is defined, the more room we have to dream up wild ideas.

As with any rule, there is a caveat to this one. Not every great invention was inspired by a problem. Sometimes, incredible inventions were created by luck or by chance or by "just fooling around." The problem solved was only discovered afterward. Chewing gum, mentioned previously, is an example of this. What prompted the inventor, Thomas Adams, to put a piece of his synthetic rubber product in his mouth? Boredom? Frustration? An instinctive desire to chew something soft and gooey? The genius here comes not in the invention, per se, but in realizing the nature of the "problem" that could be solved.

For most of us, as we proceed along the path to invent, the problem we want to solve will come first. Our challenge is to define the problem in a way that gives us a very specific and clear target to aim for but does not exclude possible solutions by its specificity. Let's look at some examples.

I sometimes give my students the following scenario and ask them to come up with solutions:

> There is a serious problem at the county zoo. It seems that the elephants are getting too many cavities in their teeth. Your team is hired by the zoo to develop a way to prevent the elephants from developing so many cavities.

The teams go to work and, inevitably, they all come up with some kind of mechanical device to brush teeth. Of course, this is a solution to the problem. But the problem statement is broad enough to allow for many other—perhaps better—solutions as well, such as fluoridating the water or changing the elephants' diet.

Had the problem been defined as "Design a toothbrush for elephants," it would be a different problem.

Here's the rub: "Design a toothbrush for elephants" is a very clear and specific problem statement. I know exactly how to proceed; the rest is engineering. Within the constraint of making a toothbrush for elephants, I can do some very inventive engineering, but in the end, it will still be a toothbrush. The problem of preventing elephants from getting cavities gives way to a much greater variety of solutions. Imagine for a moment a team of biologists, chemists, and engineers attacking this problem at the beginning of the twentieth century. Let's say they were broad-minded and looked for nonobvious solutions. Perhaps they stumbled on the relationship between diet and tooth decay. Clearly, this discovery would have had a huge impact that went well beyond elephants.

We need to define our problem with great detail and specificity in order to create a clear picture in our minds of what needs to be solved. However, we do not want to limit possible solutions in the problem statement, nor do we want to suggest the solution in the problem statement. "Design a metal clip to attach a ticket to a piece of fabric" is an engineering problem. "Design a way to attach a ticket to a piece of fabric" opens up a whole new range of possibilities. However, this could be improved by adding constraints. "Design a *low-cost* ($0.10 or less) method to *securely but nonpermanently* attach a ticket to a piece of fabric" provides a clearer picture of the problem to be solved. By using the words *method* or *way* instead of *clip*, we don't solve the problem in the problem statement, thereby limiting possibilities. By using the constraints of *low-cost* and *securely but nonpermanently*, we set up clear boundaries for our solution.

Michelangelo's masterpiece statue *David* is a prime example of using constraints to enhance a creative solution. The story goes that a statue of the biblical hero David was originally commissioned and begun twenty-five years before Michelangelo was approached. The giant block of marble was already chiseled in many areas by the original artist, who discontinued his work partway through. The partially cut block then sat exposed to the elements for a quarter of a century during which it suffered additional erosion. When Michelangelo accepted the commission to finish the statue, he used all these constraints to his advantage, fashioning his powerful image of David out of this partially hewn and damaged block. He created tension in stone—a David on the verge of confronting Goliath. Had Michelangelo started from a pristine block of marble without these constraints, would this masterpiece have come into existence?

Figure 1.3: Michelangelo's *David*. *Photo courtesy of iStockphoto.*

Is It the Right Problem?

It is worth spending time to truly understand the problem that needs to be solved before trying to generate solutions. As with a military engagement, it is essential to get "the lay of the land" before coming up with a strategy. Many times the problem you think you have turns out to be something else entirely. Sometimes, the problem you are trying to solve doesn't even exist.

In his book *Conceptual Blockbusting*, James Adams describes an attempt to design a device to retard or damp the opening of solar panels on the Mars

Mariner IV spacecraft. The solar panels were designed to open in space, where there is no air. Engineers were concerned that the force of opening—with no opposing force to damp or slow the panels—would damage the fragile solar cells. Therefore, the problem was understood to be "Develop a mechanism to retard the opening force of solar panels so they are not damaged during deployment." The engineering team created several solutions to this problem, but none proved satisfactory. Finally, with time running out and the launch imminent, the engineering team went into full panic mode. Working twenty-four hours a day at great expense, the engineers struggled to make the damper more reliable, while measuring the effects on the delicate solar panels of all the various ways the damper might fail. One of the tests assumed that the damper would fail completely and that there would be nothing to slow the opening of the solar panels. To their amazement, the engineers found that even a complete failure of the damper did not lead to an unacceptable risk of damage to the solar panels. It was only necessary to provide a shock absorber that would cushion the panels as they snapped into position. Adams concluded, "The retarders were not, in fact, necessary at all."[3]

The assumption that the solar panels would be damaged upon opening generated a faulty problem statement. Once the problem was initially stated, the engineers proceeded to seek a solution. Only the fact that a satisfactory solution eluded them led the team to discover that the problem they were solving didn't really exist.

Often, due to time or monetary pressures, problems are presented and not questioned. Whether you come up with an initial problem statement or the problem is given to you with the instruction to solve it, your job as the inventor is to pause and ask the important question: *Is this the right problem?* Other questions follow: *Yes, I see that there is a problem here, but is it the true problem? Is the problem statement too broad or too narrow? Is the problem statement too suggestive of a solution? Are the constraints well defined? Is the context understood?*

Reframing the Problem

The process of examining and restating a problem is often a greater creative act than determining the solution. As Jeff Bezos, founder of Amazon.com remarked, "The significance of an invention isn't how hard it is to copy, but how it reframes the problem in a new way."[4] Reframing a problem can lead to an entirely different perspective on how to solve it. The idea of reframing a problem is analogous to reframing a picture. When you change the frame on a picture, you view it in a different way—even though it is the same picture. When you reframe a problem, you look at it differently. For example, the difference between the problem statement "Design a bridge" and "Design a method for crossing the river" is immense. The inventor needs to spend time understanding all the dimensions of the problem he seeks to solve before beginning to contemplate solutions. Invention is the right solution to the right problem.

The Evolving Invention

Looking back in time, the obvious is very easy to see. Of course, in retrospect, the paper clip is obviously a good idea, as are the personal computer and the Declaration of Independence. But all these ideas were radical at the time of their birth—so radical, in fact, that they needed to be born in stages. The paper clip started with a much more limited and well-understood use; the personal, or "home" computer was a vision of a small group of people at a time when computers were looked at exclusively as scientific or business tools. After all, what use could a computer have besides being a niche machine for technically savvy hobbyists? Even after IBM decided there was a marketplace for personal computers worth entering, their paradigm that only hardware mattered caused them to neglect the opportunity to purchase their software provider, a tiny company called Microsoft. When the eventual signers of the Declaration of Independence first gathered to address the problem of British oppression of the colonies, they had no intention of declaring independence from Great Britain. They were merely seeking an accommodation with the ruling power. Only after repeated rejection of their attempts at compromise was the radical idea of declaring independence even contemplated.

The obvious is usually only obvious in retrospect. The great idea is rarely considered great upon first examination. Often, the power of an idea or concept cannot be captured initially by the person who originates it. A great idea may only become great as the context in which it is applied and refined changes. First word processing, then spreadsheets gave the personal computer a reason for existence. Later, the Internet made it a necessity in every home. None of these existed when the personal computer was first invented.

The inventor, much like the science fiction writer, needs to divorce himself from the real world and imagine possibilities he has never experienced. The inventor needs to learn both to open his mind to fantasy and to be the consummate realist. He is both the creative writer and the hard-nosed editor. As inventors, how do we see what will be obvious in the future? One way is to create it. During my business career, I often used Alan Kay's remark to the 1989 Stanford Computer Forum as a guidepost:

> The best way to predict the future is to invent it. This is a century in which you can be proactive about the future; you don't have to be reactive. The whole idea of having scientists and technology is that those things you can envision and describe can actually be built.[5]

While our life experience determines our vision of the possible future, today we all have a life experience that shows how rapidly fantasy can become reality. The pace of inventive and technological change over the past fifty years has been staggering. We have all experienced that. We know that when we dream up things that are not feasible today, they might be feasible, or even outdated, ten years from now. Our horizons have been expanded by our life experience. We can dare with more courage, more impunity, more imagination.

Invention in Context

Because of the rapid pace of change, we are freer to imagine future scenarios in which our inventions can function. Had I invented the automobile in the seventeenth century, would it be considered a great invention? A curiosity, certainly. Brilliant, most definitely. Intriguing in its possibilities, absolutely.

But useful?

Absolutely not.

Inventions are successful only in context. The railroad train needed an infrastructure of rails. The automobile needed an infrastructure of passable roads. They both needed an economical and practical means of supplying energy. The context of an invention is the infrastructure, technology, or cultural environment that can support the invention to make it useful. Without these, the invention remains either a curiosity or a "great idea before its time." The context of an invention determines commercial success or failure. Without wire-extruding and wire-forming machinery, the paper clip could never be mass-produced. Without mass production, our paper clip would never achieve the economy of cost necessary to make it a useful implement. While the design of the paper clip is simple, elegant, and robust, the paper clip would not have been a successful invention unless it had been developed within the context of the nineteenth-century industrial age.

Our great advantage today is the rapid change in what is possible. Figure 1.4 shows the number of patents issued per year over the past 150 years. A cursory look at the graph shows an exponential increase in patents granted.

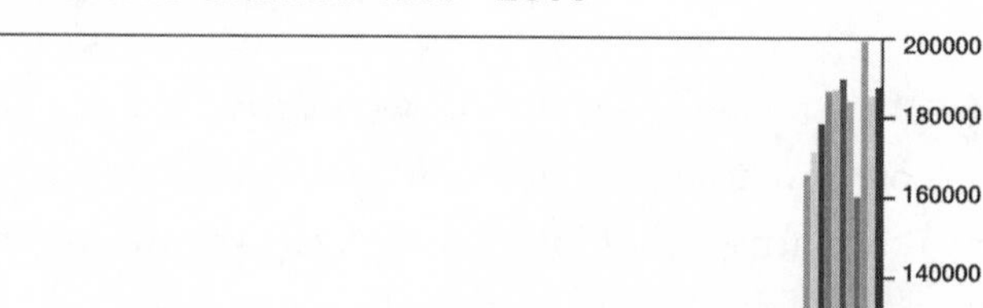

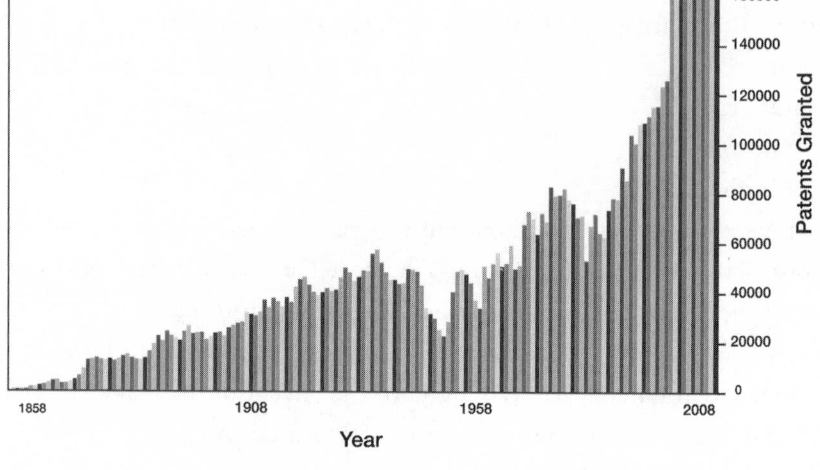

Figure 1.4: The number of US patents granted annually, 1850–2008.
Information courtesy of the US Patent Office.

This is just one measure of how rapidly our context is changing—especially in the area of technology. The area of technology has a direct effect on other areas, such as culture. We have all experienced changes in our lives wrought by the Internet. The advent of instantaneous global communication has changed how we behave as a society.

With the accelerated pace of change, new social, political, cultural, and technological challenges are being created every day. The alert inventor has many opportunities to seek out new problem areas being created by this rapid change. Change can produce many unwanted side effects in its wake. For example, while twentieth-century industrialization produced great benefits for society, it also created the by-products of air, land, and water pollution. Each of these must be dealt with in the same creative manner as the processes that brought them into existence. The twenty-first century promises vastly expanded horizons in what can be achieved. We inventors are limited only by our imaginations.

Chapter 2

THE HIDDEN OBVIOUS

THE DESTINATION

A hiker comes upon a clearing in the woods. He looks forward in the direction he wishes to travel and clearly sees the peak of a mountain. He gazes upon the peak, knowing that this is where he wants to go; it seems so close, so clearly visible, even though many miles separate the hiker from his destination. He knows that as he starts out he will reenter the woods and sometimes lose sight of his destination. He keeps a mental image of the peak in his head—he also keeps an image of the direct path to get there. In reality, his path will be indirect, winding, tortuous, and perhaps circuitous. But in his mind's eye, he can always envision the destination and the direction he must travel to get there.

The mental image of the destination drives our hiker forward. Each step along a path through the woods and underbrush is reinforced by the vision of his ultimate destination. The trail twists, descends, and ascends, and, every so often, our hiker can glimpse the peak and reassure himself that his heading is correct. He can gauge his progress.

Much of what propels him forward is belief—a clear knowledge of where he wants to go and the strong belief that he can get there. He believes both in himself and in his ability to reach his destination.

The inventor is like the hiker. In his mind's eye, he sees his destination. He believes that he possesses the wherewithal to get there. He is not sure of the exact route—he will have to feel his way—but he is certain that he will arrive. Often the invention is clearly visualized. He can see it, feel it, know it, and know that it belongs uniquely to him. He refines it in his imagination, sleeps on it, verifies its veracity in the clear light of day. He internalizes it. He owns it.

The question becomes how to get from here to there.

Being Confidently Naive

I believe that naïveté is one of the main keys to invention. Even though we can clearly see where we want to go—what it is we want to invent—we have to be naive to believe that we can really invent it. There are always huge obstacles to overcome; we cannot focus on their immensity. There are a million reasons why it cannot be done, why it will never work; we cannot abide by them or fear them. Surely, if it were possible, someone smarter than us would have done it already; we must have the courage of conviction.

In short, the inventor must be both naive and supremely confident. He must believe that he can do something that has never been done before. The courage of belief, the naïveté that he can do it, and the clear vision of what he needs to do are what the inventor must possess in order to venture forth into the unknown.

The quote below, attributed to Robert Jarvik, the inventor of the Jarvik-7 artificial heart, addresses the issue of inventors as leaders, pursuing their vision into uncharted areas:

> Leaders are visionaries with a poorly developed sense of fear and no concept of the odds against them.

One must be naive to believe that one can succeed in creating something completely novel. This naïveté is what leads the inventor to have big thoughts—ideas that stretch the boundaries of the imagination. In many ways, this is what separates the inventor from the rest of the population. This ability to think in an unrestrained way goes against the grain of our formal education. From our beginning years in school, we are instructed on how to think rationally. When we are taught to solve problems, we are shown how to proceed methodically in a stepwise fashion—to look for logical connections and progress in an orderly way. We are instructed on how to find realistic solutions that work, and we are rewarded for success. While this systematic kind of thinking is crucial to learning, it is emphasized throughout schooling to the near exclusion of any other kind of thinking.

Creativity takes a different approach. Creativity involves playing around with disorder and does not necessarily lead to immediate results. Rather, the creative process often ends in failure. Our education tells us that failure is a bad thing—something to be avoided at all costs. In order to avoid failure, we instinctively look for a solution to a problem where the path is always clearly visible. Or, we only tackle problems that are readily solvable. We are not taught that a meandering path to a solution can lead to new and unexpected ideas. We are not taught that we can learn more from failure than success. One needs to be exceedingly brave or simply naive to go beyond the learned paradigm of rational problem solving.

Many inventors who look back on a major breakthrough say, "I must have been crazy at the time to think I could actually have come up with this." This sentiment reflects the naïveté an inventor needs throughout his quest. Two other characteristics are needed to bolster naïveté: courage and curiosity. While the inventor must be naive to venture into the unexplored, he must also have both the courage of conviction and the courage to withstand self-doubt and doubt from peers. Curiosity stimulates imagination and thoughts that stretch the boundaries. The "what if?" question is one the inventor must constantly ask. This often leads him further and further from his initial ideas or preconceptions, and as one idea leads to the next, he may find himself on a completely new path. Robert Frost's poem "The Road Not Taken" beautifully expresses the inventor's dilemma of managing diverging ideas:

> And both that morning equally lay
> In leaves no step had trodden black
> Oh, I kept the first for another day!
> Yet knowing how way leads on to way,
> I doubted if I should ever come back.[1]

Like the hiker who loses sight of his destination but keeps the image of it in his head, the inventor must be willing to meander wherever his imagination takes him, while still keeping a clear picture of his larger goal in mind.

THE VISION AND THE VISIONARY

How do you get the vision? Where does it come from? This is a complex subject that I will return to in subsequent chapters. In brief, the vision can come from many places. For example:

1. **Seeing a problem that has not previously been recognized or acknowledged.** An example of such a problem is the length of time between sending a package and its delivery to the recipient. The logistics of the mail processing system dictated that the transit time was typically close to a week. Federal Express pioneered the idea that a package could be delivered the next day.
2. **Seeing a solution and discovering the problem it solves.** The invention of chewing gum described in the previous chapter is an example of having a solution and then finding the problem.
3. **Seeing a new use for existing technology.** Examples of this include the idea that electricity, when it was first harnessed, could be used for communication (as developed by Samuel Morse and others); that radar could be transformed into a cooking device (microwave oven); and that the integrated circuit could be used to make a handheld calculator.
4. **Seeing a novel solution to an existing problem.** Here the vision or insight is seeing a better way to do something that is already being done. Fasteners such as the paper clip and Velcro are examples of this approach.

Note that every example above starts with the word *seeing*. Once you can see the idea clearly in your mind, you have a place to begin. Invention often involves what I call *seeing the hidden obvious*. We all wish we were the inventor of Velcro, but it took George de Mestral going for a walk in the woods to see the hidden obvious.

De Mestral, a Swiss mountaineer and electrical engineer, went for a hike in the mountains with his dog one day in the late 1940s. There was nothing unusual in this, except that when de Mestral returned, he noticed that his pants and the fur of his dog were covered with cockleburs, the spiky seed-pods that cling to whatever moves past them so the seeds of the plant will be spread. De Mestral started removing the burrs and out of curiosity, decided to look at one of them under his microscope. He saw the surface of the burr was covered with tiny hooks which easily fastened to the fur of an animal or the clothing of a hiker. He wondered if he could somehow use this mechanism of attachment to design a new kind of fastener. He began to work on his idea by experimenting with different materials, one with a hooked side and the other with a loosely woven cloth side to receive the hooks. In due course, his experiments paid off, and in collaboration with a textile manufacturer, he eventually produced a revolutionary new temporary fastener. He called the product Velcro, a combination of the words *velour* and *crochet*.

Figure 2.1: Velcro Hook and Loop Fastener. *Photo by Ivan Boden.*

How many times have you walked through the woods and had burrs from plants stuck on your clothing? Most of us would simply consider it an annoyance. It takes an inventor's mind—a mind open to seeing in a new light that which is right in front of us—to say, "Hey, this would make a great fastener." Once that novelty is recognized, the vision is in place. Now the question becomes "How do I get from here to there?" It took George de Mestral years of hard work and persistence to make his vision a reality. But he held fast to his vision, even in the face of ridicule and temporary failure. De Mestral collaborated with a weaver employed by a French textile manufacturer to refine his invention. Eventually, he found that treatment with infrared light strengthened the nylon hooks to the desired degree to form the burr side of the fastener. De Mestral's patent, issued in 1955, enabled him to create a multimillion-dollar industry.[2]

NEEDFINDING AND PARADIGMS

Needfinding refers to looking for unmet needs in society. These needs are often not apparent until they are discovered. Entrepreneurship usually involves discovering an unfulfilled need and building a business around the solution. Needfinding is a subtle art and certainly not a science. It requires that we remove our mental filters and see our everyday world in a completely fresh light. The most difficult part of needfinding is discarding our existing paradigms.

Paradigms are mental models of the way things are. We need these mental models to get through life, to be able to anticipate the future based on a reliable model of how things happen. For example, a paradigm that I share with millions of other people is that green means go and red means stop. This mental model is a good one and helps me navigate safely in my car. According to my model, if a traffic light is green, then I can drive through it; red, I have to stop. This paradigm can be extended to other places and situations, such as industrial control rooms, where green means on, red means off. Or green means OK, red means problem. We have many paradigms that enable us to quickly evaluate a situation and engage in the correct behavior or make the correct prediction. Paradigms, because they suggest a specific way of viewing something, are essentially filters on the world. We think according to the model in our minds.

In his book *The Structure of Scientific Revolutions*,[3] Thomas Kuhn argued that these mental models, or basic assumptions, within the scientific community create a worldview that governs scientific inquiry and explanation. When information becomes available that does not fit within the paradigm, a scientific revolution occurs and the paradigm is forced to change. Kuhn calls these revolutions "paradigm shifts." Paradigm shifts are generally vigorously resisted until they are proved valid beyond a shadow of a doubt.

Paradigm shifts happen in the marketplace as well. For example, before the 1970s a timepiece involved small pieces of metal or wood moving in a circular motion around a numbered background that counted from one to twelve in a "clockwise" fashion. This was a mechanical evolution from sundials. The paradigm of a wristwatch or clock was so established that when digital timepieces came on the market, they were regarded as no more than interesting novelties. Although the shift to the acceptance of digital timepieces was slow, it is hard to imagine today that they were once thought radical. When personal computers were first developed, the existing paradigm was that there could be no real use for a small computer in the home. Computers were designed for industry or research. There were mainframes and mini-computers, and all processing was centralized. The idea of small, personal computers was far outside the paradigm. Again, it was difficult at the time for people whose worldview of computers was within the existing paradigm to imagine anything outside of it. The challenge for the inventor is to be able to step outside of his paradigm and see other possibilities.

I had the experience of being stuck in a paradigm in the early 1980s when I was a graduate student. My professor was a researcher at Xerox Research Center (PARC) in Palo Alto, California. One evening, he invited our class to the Xerox facility to see the new project he was working on. It was called the Xerox STAR system—a small computer with a revolutionary new interface. Instead of programming this computer in some arcane and technical computer language through a text editor, one interacted with the machine in a much more human and intuitive way. The software displayed images, called icons, of office paraphernalia such as file cabinets, files, and documents. Imitating real life, documents could be dragged in and out of files. This was accomplished by a small device that sat beside the computer and maneuvered a cursor around the screen. The professor called this device a mouse. (It looked a little like a mouse with its tail-like connecting wire.) The professor extolled the virtues of the STAR machine, saying that it would bring computing to the average person.

At the time, I was a full-fledged computer geek who programmed in everything from hexadecimal to FORTH. I could not imagine why anyone would want a machine that took the challenge out of computing, that made computers simpler, more understandable. "Why make computers less technical and simpler to use?" I asked. *"After all, they are not supposed to be for everyone."*

Clearly, my paradigm of computers for technical use only to be operated by highly trained people caused me to look at the STAR system with some disdain. It was a clever design, but what relevance could it have to the future of computing? Another gentleman had visited Xerox and came out with a very different impression. Where I saw an interesting novelty, he saw the future of computing. Inspired by the work at Xerox, Steve Jobs went on to develop the Apple Lisa and the Apple Macintosh.

This personal anecdote seems funny in retrospect, but the future was anything but clear and predictable back then. Jobs had a vision, but that vision was not the accepted view at the time. Like George de Mestral, he believed in his vision and had the courage to venture forth against the prevailing paradigms.

How do we get outside of our paradigms? In his address to the Stanford Computer Forum, Alan Kay quoted Marshall McLuhan: "I don't know who discovered water, but it wasn't a fish."[4] The difficulty in stepping back from your established surroundings and seeing anew is immense. However, a good start is to recognize that we are all bounded by paradigms. Only once we are aware of the limitation can we try to move beyond it.

A key to needfinding is getting outside of our paradigms. In the last chapter, I discussed the "elephants with cavities" problem. The engineering paradigm says, "Develop a toothbrush"; someone who is outside of that paradigm might say, "Change the diet." If you think this is semantics, take a brief look at paradigms in modern medicine. With the development of commercial antibiotics in the 1940s, the paradigm in medicine became "cure problems with these miracle drugs." As the pharmaceutical industry developed, the paradigm was modified to "cure every problem with drugs." Nathan Pritikin's suggestion in the early 1960s that a low-fat, low-cholesterol diet could be a key to preventing heart disease was laughed at by the medical community.[5] If the cure (people weren't thinking about prevention then) couldn't come from the latest drug technology or surgery, so the paradigm went, it could not exist.

As inventors surpass the limitations of their paradigms, needfinding becomes an exercise in keeping one's eyes open. Most often, both needs and solutions are right in front of us. We just need to be able to see them. George de Mestral was able to reframe a common nuisance experience into a vision for a new fastener. In retrospect, his discovery seems obvious: a "Why didn't I think of that?" kind of discovery.

Seeing the Hidden Obvious

There are needs that broadcast themselves for everyone to see. For example, there is a need to develop new antibiotics that solve the problem of drug-resistant bacteria. Solving this problem will be both financially rewarding and a great contribution to society. This kind of need requires tremendous resources to find a solution, and the solution is often based on inventing a new technology. The *hidden obvious* needs, such as de Mestral's, are different. Their discovery requires a unique combination of the following abilities and traits.

1. Active awareness
2. Technological fluency
3. Seeing the flower from the seed
4. Desire to create, curiosity
5. Courage

1. **Active awareness.** You have to keep your eyes open and be on the lookout. We swim in an ocean of information. This information passes

through us via all our senses. For the most part, we try to swim with the current, only making use of the information that we concretely need. Due to the quantity of information or stimuli we are presented with every waking minute, we need to be able to filter out most of it. Certain mental illnesses are a result of not being able to filter out external information. However, this filtering gives us a narrow focus. Active awareness means being aware that there are things we experience that, when seen in a different context, can create a new solution to an old problem. Sometimes this process is called refocusing. A good example of refocusing, or seeing something in a different context, is the examination of visual illusions, such as the one shown in figure 2.2. Does the figure show a young woman or an old woman? Can you see both?

Figure 2.2: Old/young woman.

(Hint: one is looking toward you, the other looking away). If you have fixated on one of the images, it is hard to see the other without squinting or somehow changing your perspective. Active awareness is openness to meaning other than the obvious literal one that you see first.

Friedrich August Kekulé, the scientist who discovered the shape of the benzene molecule, studied the nature of carbon bonds for years and still could not reconcile the properties of benzene with what was known about carbon bonds. After working on the problem one night, he fell asleep in his chair and had a dream of a snake chasing its own tail. He awoke with a start and had the insight to see the dream as a solution to the structure of benzene.

> But look! What was that? One of the snakes had seized hold of its own tail, and the form whirled mockingly before my eyes. As if by a flash of lightning, I woke; . . . I spent the rest of the night working out the consequences of the hypothesis. Let us dream, gentlemen, and perhaps we shall learn the truth.[6]

Kekulé could have interpreted his dream to mean many things, or he could have simply ignored it. If we consider that this problem was foremost in his mind and that he had a discoverer's awareness that things in his everyday world might contribute to a solution, we can understand how he could see in this bizarre dream the solution to the problem he was working on in the laboratory.

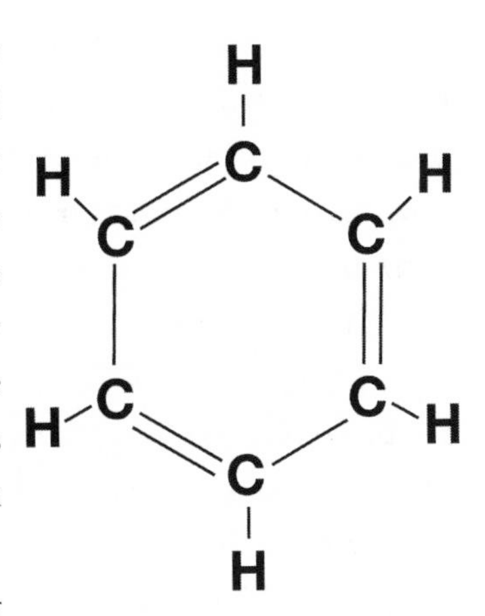

Figure 2.3: The chemical structure of benzene.

Deciphering significant new meaning from ordinary information—whether it be plant-seed burrs stuck to clothing or a vivid nonsensical dream—requires an active awareness that all information might have significance. This becomes a state of mind, a state of readiness, a willingness to take a second look at things and see what they have to say to you.

2. **Technological fluency.** In order to be able to recognize need, you must have an understanding of your area of investigation. You need to be fluent in the technological "language" that describes the area. For example, if my area of interest was reducing wasted energy, I would need to be conversant with the technologies of energy production, conversion, and use. With this background, I would be able to look around and understand where needs in this area exist. I would have the knowledge needed to both recognize a new need or discovery and to evaluate it.

Technological fluency is different from expertise. Fluency means being conversant in the technology, knowing enough to be able to recognize something that is significant, and knowing where to look for a more in-depth understanding. De Mestral was not a textile expert, but he was fluent enough in the technology to know the direction to take and how to gain the expertise needed. He was also fluent in technological evaluation—a key for any inventor. All ideas must be evaluated to see if they have merit. I don't know whether the story of de Mestral's epiphany at his microscope is accurate or apocryphal, but it shows the "aha!" moment and the use of technology—in this case, a microscope—to verify a hunch. De Mestral had an engineering background and used his general knowledge as a springboard to delve into the specific area of his invention.

Technological fluency is most often gained by a study of the area of interest. This includes reading periodicals, electronic media, formal study of basic principles, attending trade shows, lectures, and so on. The goal is to become conversant enough in your area of interest that you will be able to recognize new needs and possible solutions. Once you are hot on the trail of something, you can focus your attention on gaining expertise.

3. **Seeing the flower from the seed.** As I described in the introduction to this book, I used to travel with my father on needfinding expeditions. We would go to trade shows in a variety of nonrelated areas and walk the aisles of the show just looking for interesting things. This was my apprenticeship to needfinding. Most often, my father had no idea what he was looking for. He would usually just observe, see things, and catalog them in his mind for use later on—sometimes years later. Often the seed of a need or future invention was there, and it simply needed to be developed and placed in context. An example of this was an actuation and nozzle mechanism shown at a trade show for dental equipment. No doubt this was designed for the controlled release of a jet of water. The mechanism intrigued my father because of its simplicity and elegance of design. It was eventually incorporated into a product developed by my father called the MicroDuster, a can of compressed gas used for cleaning in critical environments. Was the elegant nozzle the seed of this invention? Or did it sit in the back of his mind to be recalled as the perfect solution to the metering of his compressed gas product? Either way, the seed—which came from a completely different area—was brought to flower.

4. **Desire to create.** Inventing is the act of creation. Creation is never an act of comfort. There is no well-worn path to blindly and automatically follow. In order to tolerate the discomfort of creating something, you need

to have the *desire* to create. Sometimes this is referred to as having curiosity, but it is more than that. Curiosity does not always spur us to action; desire does. Desire implies a drive toward attainment. As previously discussed, the inventor is one who blazes a new path. This is never easy and is often fraught with setbacks along the way. There are always reasons to give up. The desire to create, the desire for fulfillment through creation, is a driving force for every inventor. We need to follow this desire and suspend disbelief in ourselves and our abilities to create something worthwhile and new.

5. **Courage.** Courage is the force behind desire. Yes, we want to invent, we have the desire, but we also need the courage to act. We need the courage to continue in the light of failure. We need the courage to believe in ourselves and our ideas when no one else does. We need the courage to ignore all the reasons why the need isn't really significant or the solution will never work. We need the courage to be alone. We need the courage to keep our vision alive and illuminated in our mind. We need the courage to continuously nurture our idea. Thomas Edison addressed the question of courage when he said, "Many of life's failures are people who did not realize how close they were to success when they gave up." When he said that invention is "98 percent perspiration," he was referring to trial and error, dealing with failure, and the overcoming of self-doubt that challenges the inventor at every turn. The courage to continue slogging along is essential.

Successful inventors by nature have tremendous courage. From their initial insight to the actualization of their discovery often takes years of effort. De Mestral's more than ten years of toil to perfect Velcro is an example of the courage it takes to continue. Another incredible example is that of Chester Carlson, who invented xerography—arguably one of the most significant inventions of the twentieth century. It took Carlson over a year to go from his idea, which was described in a patent application to an actual reduction-to-practice (working prototype or demonstration). He then pursued his idea for twenty-two years before it actually reached the marketplace! The courage to persevere for twenty-two years is something most of us could not imagine.

The following is from an interview with Bob Gundlach, Xerox's first research fellow, from the book *Inventors at Work*:

> After his patent application, he then worked on it unsuccessfully in his own kitchen and bathroom for about a year. Finally, he hired a helper, Otto Kornei, who was able to help him get a reduction-to-practice within three weeks. The first xerographic image was of the place and date 10-22-38 [October 22, 1938, in Astoria, Queens, where he was living at the time]. . . . It was a small image, maybe a little less than two inches long, but it represented the copying process as we know it today.
>
> When he started, he had no idea that it would take twenty-two years to get it to market. He approached at least twenty major business equipment corporations in America and received what he later termed "an enthusiastic lack of interest" from all twenty—as well as from the National Inventors Council and other federal agencies.[7]

Eventually, Carlson made contact with Haloid Company, which was searching for a new product direction. Haloid was the former name of Xerox Corporation. The first Xerox automated copier—the Haloid Xerox 914—was introduced in 1960. We will return to the Xerox story in chapter 9.

With something new, especially something revolutionary, there is often no established market, nor is there a way to predict commercial success. Often there are many failed attempts, both at perfecting the invention and making it commercially viable. Success requires great persistence and courage. Once the inventor sees the hidden obvious and has his insight, he needs the courage to continue to believe in his idea in order to bring it to fruition.

The Vision Refined

Once you have the vision, how do you refine it? How do you develop it and give it clarity? A good way to start—especially if the need or idea that you envision involves areas or technologies that are unfamiliar to you—is to gain *contiguous expertise*. Contiguous expertise means studying subject matter in all areas that in some way touch upon your area of focus. For example, if I were to take de Mestral's walk in the woods and had a vision of a new kind of nonpermanent fastener based on plant burrs and fabric, I would delve into the areas of fasteners, botany, textiles, and flexible materials. Not that the answers to all my questions would lie in textbooks, but immersing myself in the areas surrounding my discovery would stimulate my brain to make connections that would otherwise never be made. Often, invention comes from a connection made between two completely unrelated areas.

A second, and sometimes neglected, area is that of documentation. An inventor needs to record his process. Keeping a detailed record of your thoughts, ideas, and research is essential to solidifying your thinking. Documenting the exploration and refinement of ideas with diagrams and written descriptions goes back a long way. Leonardo da Vinci's sketchbooks are a wonderful example of combining visual thinking with textual ideas.

Bob McKim, in his book *Experiences in Visual Thinking*,[8] describes the "idea log" or "design log" as a way to record your ideas. He suggests that the form can be anything from index cards to a notebook to a long scroll of paper that, due to its lack of boundaries, encourages the flow of ideas. He stresses drawing as a way to express ideas with greater rapidity and fluidity than textual descriptions. This gives you the freedom to think with your hands. As you refine your visions, it is important to understand that you are constantly switching between the analytical and the creative. A design log is a means to combine the two. Pages of notes can be analyzed and key ideas seized and expanded upon in quick sketches. The idea behind the design log is that it shows movement of thought. Thoughts are generally not neat and ordered—especially creative thoughts. The idea log should represent the process of thinking with all its dynamism and sloppiness. You research, think, make connections, and create—all on the paper in front of you. This eventually becomes a document of your process, much like Leonardo's sketchbooks. The log, however, is not meant to be a finished work of art. It is a tool to help you invent. It reflects your thought

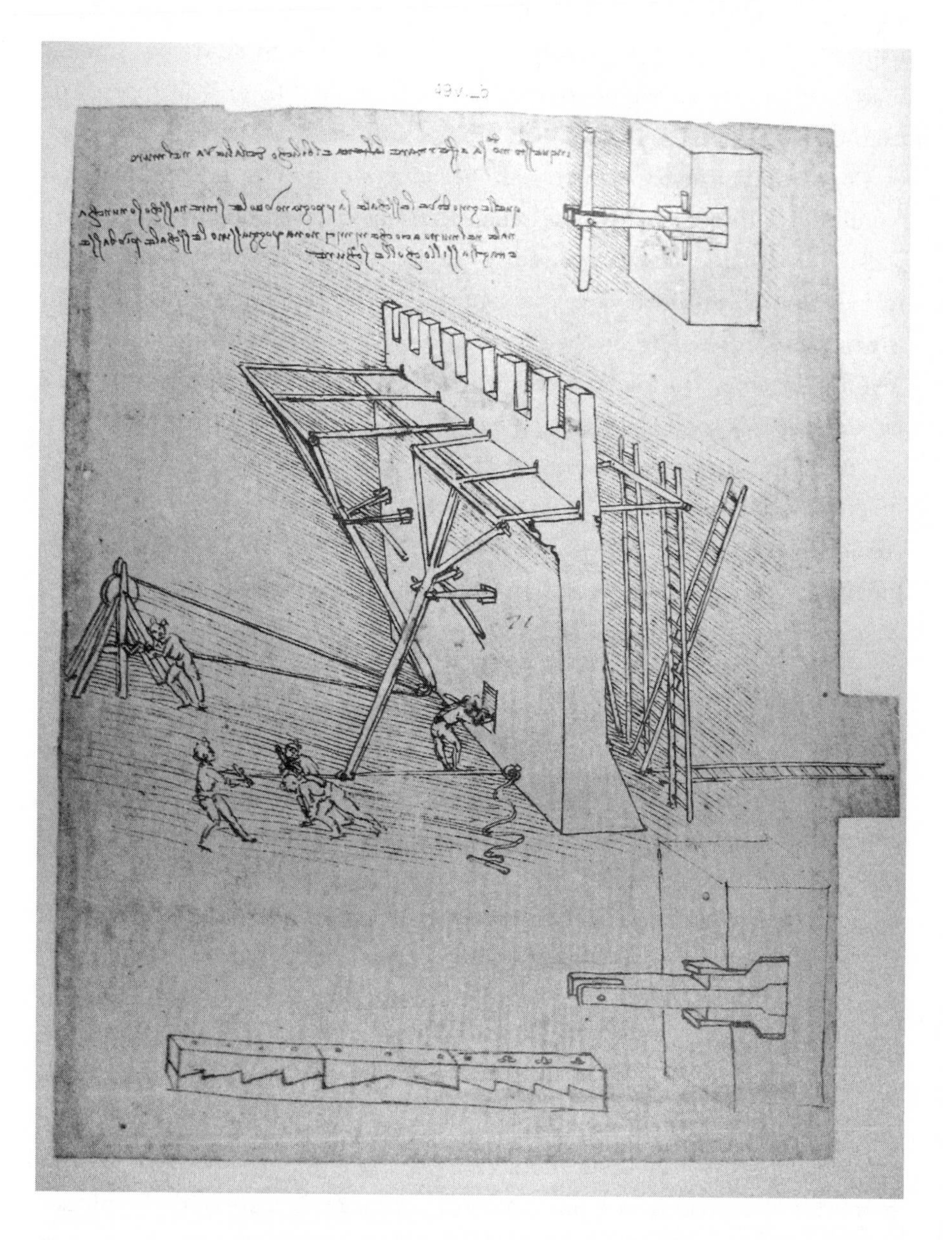

Figure 2.4: Leonardo da Vinci's sketch of a moving wall to defend against attackers. *Photo from Freerangestock.com.*

process and thus will have a more "wild and scribbled" look to it. Figure 2.5 shows several pages from a design log I created for a certain project. I had a general vision of what I was trying to do and used the log to refine the vision. The log explores multiple ideas and directions in quickly sketched drawings with some textual notation. The check marks, inserted at a later time, represent a review and consolidation of the ideas.

The idea behind visual thinking is that you employ your senses in a more varied way than simply writing down words. Both the kinesthetic feedback that comes from drawing and the visual feedback that comes from seeing a drawing stimulate the brain in more ways than a conventionally written document. As inventors, our brains need all the stimulation they can get. It only makes sense to involve all the senses. We can add to this by including the tactile sensation. Many people are most creative when they can get their hands around something. The design log can be expanded to include actual physical models. For example, if you are trying to come up with a mechanism or structural design, toys like Erector Sets or LEGOs can be of great assistance. Sometimes simply working clay or a similar formable substance with your hands can help generate new solutions. Models can be inserted into a design log simply by taking photographs and pasting them in. This provides a record of the creative action and will remind you of the process. Digital cameras make this very easy. Software readily available for the personal computer can add yet another dimension to the integration and manipulation of text and images.

Once we have somewhat refined our vision, we need to challenge it. Challenging our vision involves examining alternatives, playing with the "what if's." For example, let's say my vision was the design of a vehicle that would go very fast on smooth surfaces and yet would have the capability of driving over rugged terrain as well as densely forested areas. I envision a wheeled vehicle with retractable legs that can extend from its chassis when entering rough or forested areas. I have the design roughed out on paper and the unique mechanisms pictured in my mind. Now let's challenge all this. What if I were to use a single method of propulsion for all cases: for instance, forced air used to propel the vehicle both horizontally (forward motion) and vertically (up-and-down motion)?

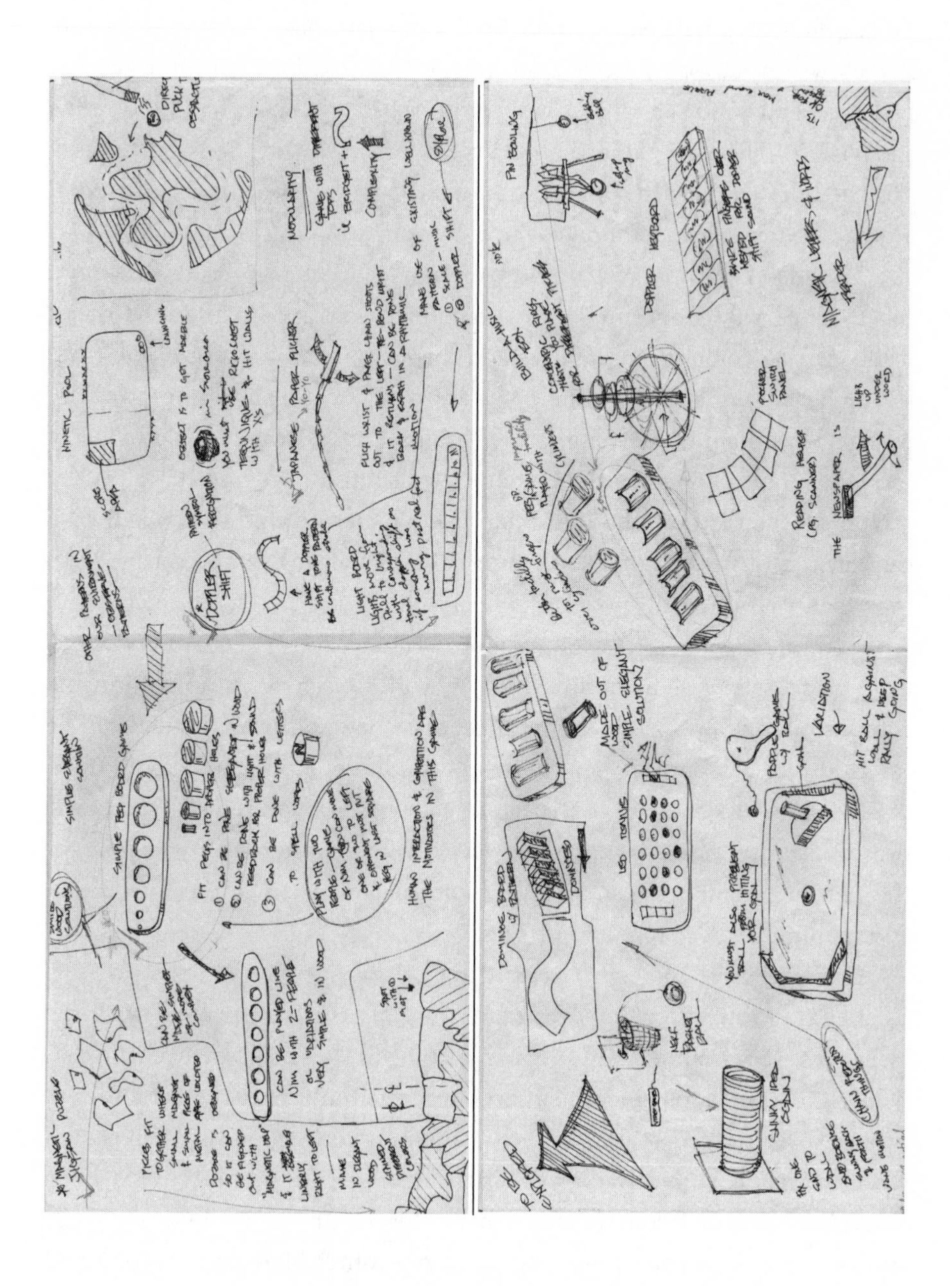

Figures 2.5: Pages from Author's Design Log.

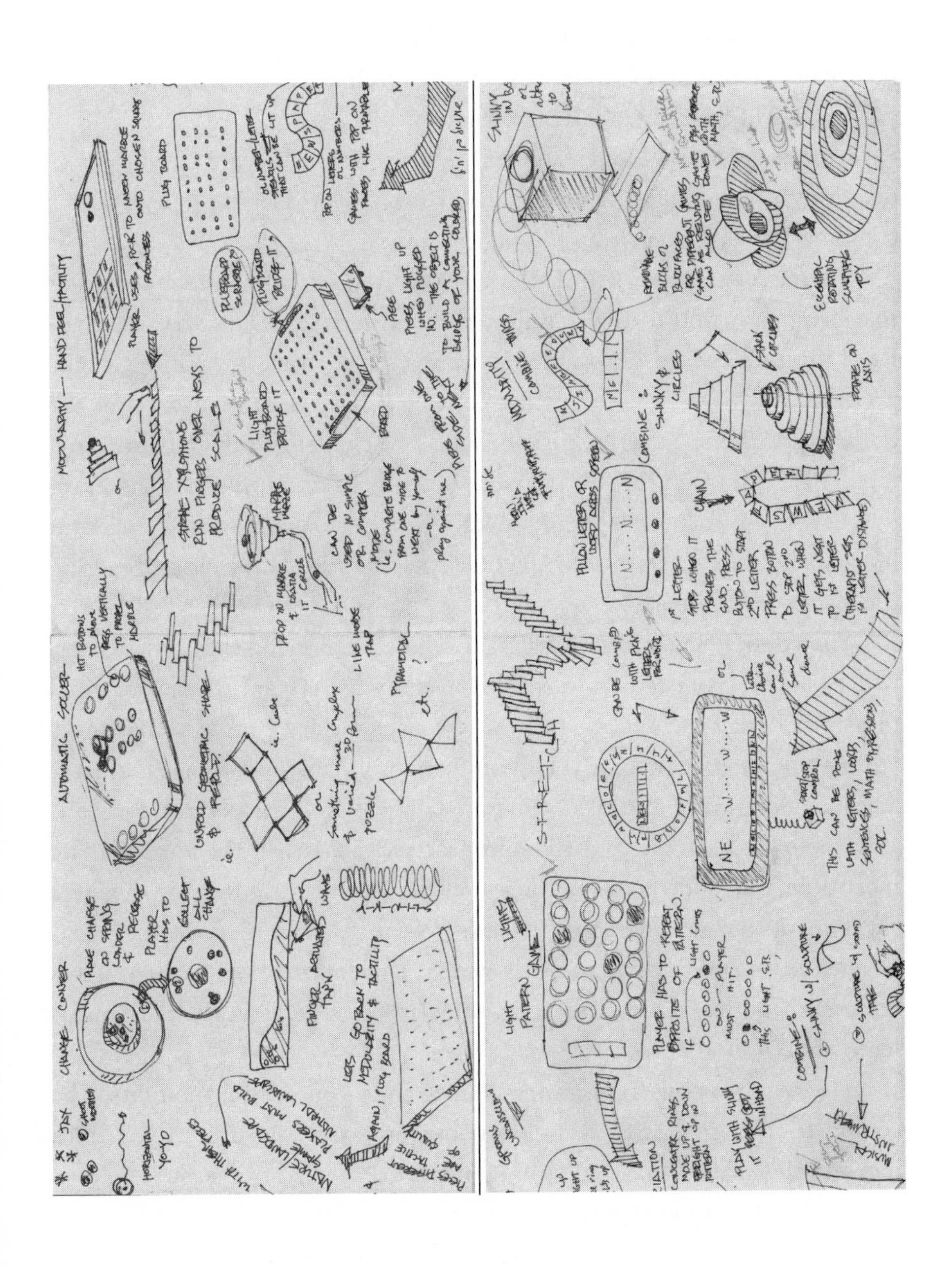
AUTOMATIC SOLVER
PLUG BOARD
LIGHT PATTERN GAME
HORIZONTAL YOYO
MUSICAL INSTRUMENTS
ROTATE ON AXIS

Imagine this and compare: Is it feasible? Is it more efficient? More maneuverable? Simpler? More cost-effective? Or what about spring-loaded wheels or wheels on hydraulic lifts that act like legs?

Challenge the problem statement. Does the vehicle really need to drive over rough terrain? Can it be designed to skip over terrain by making it a combination land vehicle and flying machine?

There are many ways to challenge the initial vision, and each one leads to a sharpening and further refining of the idea. It's possible that the challenge might lead to a completely new idea or way of looking at the problem—perhaps something that could not have been seen without the exercise of delving into the initial vision. The challenge can be frustrating, especially if you find that all the work you did on refining your vision produced a vision with some gaping holes. The work was not wasted. Without it, you would never have seen the holes or the new ideas that challenging the vision can present.

Your design log can be used as a take-off point for these challenges, which can be expressed and explored visually in relation to your original ideas. Challenge leads to the process of consolidation. Once the challenges are explored and answered, you begin to consolidate your ideas. The great advantage of having all this documented on paper is that you can review it as a whole. You can see your thought processes and select the ideas that are most promising. In the design log sections shown in figure 2.5, there are check marks on certain pages. These represent the culling of the ideas that I eventually wanted to pursue. Sometimes, none of the ideas by themselves represent the final vision, but a combination of several suggests the ultimate direction.

The vision is now significantly refined from your initial thoughts. The combination of research, creative documentation in the form of your idea log, and the challenging and culling of your best ideas has led to a refined and sharpened vision. If you are satisfied, great! If you still don't think you are there, this process can be repeated. The repetition of this entire refinement cycle can be very effective in revealing the essence of your vision with a clarity that it did not have when you started the process.

Chapter 3

CREATIVITY AND THE BRAIN

"The mind is an iceberg—
it floats with one-seventh of its bulk above water."
—Sigmund Freud[1]

THE CREATIVE MIND

We are not educated to be creative. We are educated to use our rational, conscious minds to solve problems. School systems throughout the world follow curricula that teach rational problem solving. This is an important thing to do; without the ability to analyze and develop rational approaches, the citizenry would not be able to function in today's industrial, technological world. Businesses would not be able to overcome the day-to-day challenges necessary to move forward. The conscious mind is primed to gather, assess, and act on information in a purposeful way. We are taught to connect the dots, to make logical sequences, to draw sensible conclusions based on analysis of information. We are taught in this manner as soon as we are educable.

Without these essential skills, even the most creative person cannot be productive. However, creativity is usually sacrificed in the effort to master rational problem solving. Creativity and rationality stem from different sources. Creativity is neither linear nor logical. It is often wild and irrational. It does not derive ideas in logical sequence but makes large leaps that are seemingly unconnected. It is not the product of a conscious effort but of something beyond our deliberate control.

Psychologists divide the mind into the conscious and the subconscious.

The conscious mind deals with those things that you are aware of and can manipulate. The subconscious mind deals with deep memory, the autonomic nervous system, and the ordering of the vast quantity of information your senses take in. You are generally unaware of what goes on in your subconscious mind. For example, breathing is a subconscious function; you are not aware of every breath you take. However, this is something you can easily bring into your conscious mind by focusing your awareness on it. You can count your breaths, but if you get tired of counting or move to another area of awareness, you will not stop breathing. Your subconscious will keep the machine going. For our purposes, we can think of awareness as the difference between the conscious and the subconscious. It is estimated that the subconscious mind employs approximately 80 percent of the brain's neural capacity. The subconscious mind processes a tremendous amount of information with only a small portion being brought into your awareness. This ability to process and organize a large amount of information is the basis for creativity. When we are creative, we often make connections between seemingly unrelated ideas or pieces of information. There is no sequential stepping in logical order from one piece to the next but rather a huge jump. Only our subconscious can give us these leaps. I sometimes imagine the subconscious to be a boiling cauldron of information with ideas shooting up into the air like streams of vapor, only to re-condense and re-form with other material and boil off again. The creative process depends on capturing the ideas and connections that you need and implanting them into your conscious mind.

This somewhat mysterious process involves four mental activities:

1. Feeding the brain by accumulating the information that will serve as the building blocks of creativity,
2. Listening to the subconscious as it works with the problem,
3. Using intuition, or thinking with the gut, and
4. Evolving an idea using your inner senses.

FEEDING THE BRAIN

In order to prime our subconscious to work on creative connections for us, we need to give it information to work with. In the preceding chapter, I spoke about attacking a problem or fleshing out an idea by developing contiguous expertise in the areas surrounding the idea or problem. This means doing extensive research and flooding your brain with information related to the area you are focusing on. The effect of doing this research is to feed the brain with information about your subject. You do not have to consciously determine which pieces of information are valuable; your subconscious mind will sort though the mass of data. Your job is to feed it.

Have you ever experienced the frustration of concentrating on a problem and not being able to solve it? Then, after giving up, you awake the next morning with the answer clearly in your mind. Most of us have had this experience. This is how the subconscious works. Your attention to the problem and your search for a solution gave the mind the food it needed for processing. The subconscious processed the information and provided the solution upon waking. The famous story of Kekulé's dream, described in the previous chapter, is a perfect example of this. Unfortunately, the subconscious does not provide solutions on demand. It works very differently from the conscious mind. We need to be both patient and attuned, so we hear it when it speaks to us.

Two inventors interviewed in Kenneth Brown's book *Inventors at Work* speak about this process. Jerome Lemelson, who holds over four hundred patents in a variety of areas from toys to integrated circuit manufacturing systems, has the following to say:

> On a number of occasions, I've woken up in the middle of the night with the solution to a problem that has been on my mind. Sometimes I've also come up with ideas that were totally unrelated to anything I had ever done. There it would be—either the solution to a problem or the heart of the problem itself.
>
> . . .

> The mind has to think about a problem subconsciously. We still don't know what goes on in the subconscious mind. There are many processes of short-circuiting, elimination, and so on at work. I'm sure these processes go on subconsciously. But there is a combination of things at work behind creative thoughts.
>
> . . .
>
> . . . you may forget about it [a problem] for a while and the solution may come to you while you are working on something else. That's happened many times to me.[2]

Stanford Ovshinsky, the self-taught physicist and inventor who revolutionized the use of amorphous materials in the semiconductor industry, described a similar experience:

> I have to go through what I call my war dance to invent. First, I have to know what the problem is. Then, I do wide-range reading about the problem. And I have to work very hard thinking about it. . . . Once that's done, I can do anything else—I can be at the beach, I can be walking in the woods—and suddenly, I will get the answers I want. Then, I hurry back to my workbench and try them out.[3]

Lemelson and Ovshinsky describe exactly the same process:

1. Define the problem,
2. Flood your brain with information through intensive research, and then
3. Let your subconscious mind go to work.

Creative individuals have a fine-tuned sensitivity to their subconscious and a developed awareness of how to listen to it.

Gordon Gould, interviewed in the same book, describes the flash of insight that led to his invention of the laser. The flash, he says, would not have been possible without the years of work leading up to it.

> In the middle of one Saturday night in the fall of 1957, the whole thing . . . suddenly popped into my head, and I saw how to build a laser. . . . But that flash of insight required the twenty years of work I had done in physics and optics to put all the "bricks" of the invention in there. The process of invention is fascinating in any case, and it's a process that is not fully understood by anybody, least of all me. But I have learned that it is necessary to have all the materials of an invention in your head. . . . I think the mind is unconsciously churning away, putting all these things together like a jigsaw puzzle. . . . Every once in a while, something really clicks, and an idea will spring into your mind. It only seems that it was instantaneous.[4]

Listening to Your Subconscious

"The conscious mind may be compared to a fountain playing in the sun and falling back into the great subterranean pool of subconscious from which it rises."
—Sigmund Freud[5]

"Man's task is to become conscious of the contents that press upward from the unconscious."
—Carl Jung[6]

How do we learn to use our subconscious effectively? As I write this, I think of a very recent experience. I was working on a complex robotics problem that involved navigation with a mobile robot. One of the annoying glitches was the tendency of the robot's wheels to get caught when it was moving parallel to an irregular wall or turning a corner. I also wanted to keep track of the distance the robot traveled parallel to a wall. I left this problem unsolved for the day and went about my other business, which had nothing to do with robotics. That night, just before I fell asleep, the solution popped into my head. Of course! It's simple! Add another wheel that would rotate perpendicular to the drive wheel. This wheel would both protect the robot from getting stuck and enable me to measure the distance traveled. Perhaps someone else would have seen this solution—or an even better one—right away. But I did not see a solution at the time. While I consciously forgot about the problem, my subconscious went to work. And when it was ready, the answer popped out. Now I am free to pursue this direction and, perhaps, using conscious thought, improve upon it.

Using our subconscious effectively requires patience, trust, and recognition. There are things we can do to get the subconscious working—such as the aforementioned intensive research into our problem area. Once we do this, we need to trust the subconscious to produce a result and have the patience to let it work in its own time. Most often, the boundary between our subconscious and conscious thought is most permeable when we are in an extremely relaxed, semi-dreamy state. Such a state occurs naturally right before sleep or upon waking. At these times, rational thought does not dominate our minds, and we are better prepared to access the subconscious. We can purposely put ourselves in such a state through relaxation exercises such as meditation or conscious body relaxation. Relaxation techniques are typically promoted to improve health and reduce stress, and they can also help attune your conscious mind to the workings of your subconscious. Such techniques include self-hypnosis, guided imagery, yoga, hydrotherapy, and biofeedback. Bob McKim, in his book *Experiences in Visual Thinking*, refers to this state as relaxed attention.

> The importance of relaxed attention to creative thinking is well known. After intensive conscious preparation, the creative thinker commonly lets the problem "incubate" subconsciously: "I will regularly work on a problem into the evening and until I am tired. The moment my head touches the pillow, I fall asleep with the problem unsolved." After a period of relaxed incubation, which can take place in the shower or on a peaceful walk as well as sleep, attention is not uncommonly riveted by the "aha!" of sudden discovery. "Frequently I will awaken four or five hours later . . . with a new assembly of the material." While subconscious incubation requires relaxation, a sudden flash of insight requires attention or it is lost. Again, relaxed attention.[7]

Intense exercise, while not relaxing in the least, causes the brain to produce chemicals called endorphins, which produce a state of euphoria during exercise. Joggers refer to this state as "runners' high." The effects of endorphins also include a sense of calm and serenity that lasts well beyond the exercise period. This post-exercise state is one in which you may also become more aware of messages from your subconscious.

Scientists have measured electrical activity in the brain and divide the types of activity into four kinds of brain waves—alpha, beta, delta, and theta. Beta waves predominate during periods of intense concentration or rational problem solving. They are also common during stressful periods. Delta waves are produced during sleep, and theta waves indicate a deep meditative state. Alpha waves are produced during a state of alert relaxation. For our purposes, we would like to be able to go into an "alpha state" to foster our creative potential. One popular method of attempting to control brainwaves is biofeedback. Biofeedback involves monitoring both brainwaves and autonomic nervous functions such as pulse, body temperature, skin moisture, and respiration while providing a visual readout of the state of these parameters. The person undergoing biofeedback consciously uses relaxation exercises to try to control these functions. She can monitor success as she experiments with various techniques. Biofeedback is in essence a tool to teach relaxation. Through the use of external feedback, the user learns how to control her autonomic nervous system and produce an alpha state of alert relaxation. Once the technique is learned, the external device is no longer necessary. There are many biofeedback devices and programs sold in the marketplace today. Often, these devices hook up to a personal computer and can be used in the home. Software in the form of games or visual challenges is used to provide the feedback in order to make the process more interesting.

In this predominantly alpha state, you are relaxed but also attentive enough to be aware of ideas that flow from your subconscious. When a good idea becomes apparent, you must be attentive enough to grab it so it is not lost.

Inventors spend a good deal of time in an alpha state. This generates the stereotype of the dreamy or absentminded inventor. Trying to optimize idea flow from the subconscious into the aware mind is key to the inventor's creativity, and this lends an aura of truth to the stereotype.

THINKING WITH THE GUT

Another aspect of nonrational thought is what is commonly known as intuition or "gut feeling." For the inventor, this is an important avenue to harness creativity. In fact, it often serves as a bridge between spontaneous creativity and rational thought. Our gut feeling can lead us in a direction or give us a reading on an idea or train of thought. Inventors often act on "hunches" as to where to take their next steps. Intuition is usually thought of as ephemeral and hard to pin down. It is not always reliable and cannot always be called upon. However, it is very powerful.

What is gut feeling or intuition and why is it so powerful? It is a kind of internal knowing that provides guidance when the outside information is unclear. In his book *The Second Brain*, Dr. Michael Gershon describes the gut as having neurological capability independent of the brain.

> The enteric nervous system can, when it chooses, process data its sensory receptors pick up all by themselves, and it can act on the basis of those data to activate a set of effectors that it alone controls. The enteric nervous system is thus not a slave of the brain but a contrarian, independent spirit in the nervous organization of the body. It is a rebel, the only element of the peripheral nervous system that can elect *not* to do the bidding of the brain or spinal cord.[8]

He speaks in reference to gastrointestinal disease, but the metaphor of the gut as a second brain has long been with us. Intuition represents a different kind of thinking. In fact, it is not thinking but feeling. We feel certain things that we can't quite put our finger on, and yet the feeling is strong and pervasive. Can we trust these feelings? Can we harness them? Do they have validity?

I would argue that gut feeling is a key quality not only in creativity but also in leadership. Leaders, whether in business, government, or other fields, know how to harness these feelings and follow them, especially in times of crisis. Inventors and those engaged in creative projects also learn how to depend on their gut feelings to make decisions. The hardest part of learning to harness intuition is learning to trust it. In the absence of rational reasons or outside verification, it is difficult to trust a feeling. People who have mastered this—who have learned to trust their intuition—have learned to do so through experience. They have had the experience of intuition being correct in the face of neutral or opposite information coming from rational thought. As these experiences multiply, one learns that intuition can be a very powerful ally in decision making.

Gut feeling is not always correct, nor can you always summon it up. But it is something that must be nurtured and paid close attention to in the inventing process. It often leads us to places where we would not otherwise go. Coupled with information coming from our subconscious, gut feeling can guide us in filtering ideas and choosing the correct direction.

YOUR INNER SENSES

"I need to touch music as well as to think it, which is why I have always lived next to a piano."
—Igor Stravinsky[9]

Just as we perceive the world through our five senses, we can perceive ideas through the same senses internally. A composer hears music in his mind; a chef can mentally experiment with recipes as he conjures up the taste of ingredients; a dancer can rehearse a routine in her mind, both visualizing and feeling movement. When it comes to inner sensing, most people tend to favor one or two senses. They relate to abstract concepts by imagining them with their strongest inner sense. For example, my strongest inner sense is kinesthetic. I like to *feel* ideas. Touching parts or mechanisms when trying to come up with a mechanical solution is very helpful to me in imagining what the solution will feel like. My secondary inner sense is visualization. Combining the two enhances my ability to imagine and create. Building a model or a prototype is a way to stimulate both of these inner senses. Often we think of prototyping as a way to show others our idea, but I see it as a way to stimulate mental development of the idea. It's not the prototype itself that is most important; it is the act of making the prototype that leads to fresh insight.

Others solely use visualization. They see a problem in their "mind's eye" and work it from there. People who are very strong in this area can actually run processes or create designs completely in their minds. Temple Grandin, a professor at Colorado State University, is one of the world's foremost experts in designing slaughterhouses for cattle. Her designs are praised as humane in the sense that, while they work with the efficiency of a mass-production facility, they keep the animals calm, comfortable, and free of pain throughout the process. Temple Grandin is also autistic. An effect of her autism is that she can see and manipulate complex engineering designs in her mind, without drawing them on paper. She describes this ability in her book *Thinking in Pictures*:

> Every design problem I've ever solved started with my ability to visualize and see the world in pictures. . . . Now, in my work, before I attempt any construction, I test-run the equipment in my imagination. I visualize my designs being used in every possible situation, with different sizes and breeds of cattle and in different weather conditions. . . . When I do an equipment simulation in my imagination or work on an engineering problem, it is like seeing it in a videotape in my mind. I can view it from any angle, placing myself above or below equipment and rotating it at the same time. . . . I can visualize the operation of such things as squeeze chutes, truck loading ramps, and all different types of livestock equipment. The more I actually work with cattle and operate equipment, the stronger my visual memories become.[10]

Memorization is a common activity that requires the use of inner senses. Think of how you go about memorizing something: perhaps you hear it mentally or see it written out in your mind. In sports, coaches encourage the use of inner senses to improve performance. Often this involves stepping through the athletic activity in your mind and doing it perfectly. Perhaps you imagine the feel of your muscles tightening or extending with each step. The assumption is that the mental experience will translate into actual performance. I recall doing this when I learned to fly an airplane. I was learning instrument approaches to landing, which are complex and involve keeping track of many different things at the same time. I would mentally rehearse the approach sitting at my desk with my eyes closed. I would move my hands as if they were on the yoke and throttle and visualize in my mind each step that needed to be taken. This helped me master the complex choreography and situational awareness of an instrument approach.

The inner senses require a state of relaxed attention with no outside interference. Unlike receiving ideas from our subconscious mind, this is an active process that we direct. We consciously manipulate ideas in our imagination and need undisturbed concentration in order to keep track of their flow and development. *Imagineering*, as this is sometimes called, requires that the inventor establish an environment where imagination can flourish. This might be a comfortable chair in a quiet room or a workbench surrounded by partially built mechanisms, or it might be a couch where one can lie while listening to music. There are as many optimal environments as there are creative people. The task is to find what is ideal for you and create that special space where you can indulge your inner senses.

Working Ideas, Percolation, and Not Falling in Love

We have looked at various ways to enhance creativity—through monitoring subconscious thought, intuition or gut feeling, and conscious direction of our inner senses. These are like instruments in the orchestra of our creative mind. They all work together to produce and refine ideas. The concept of working an idea and letting it percolate has to do with using multiple avenues to refine creative thought. When I think of the word *percolation*, I think of an old-style coffeemaker in which the water boils up through a tube and filters back through the coffee grounds. The process happens over and over until the coffee is ready. Our minds automatically percolate thoughts. We inject a problem into our minds by thinking about it. It then automatically percolates through our thought process.

Working an idea means consciously thinking about it and looking at various permutations. Working an idea usually does not mean coming to a conclusion or a solution. You play with the idea, trying different variations on for size. You manipulate and recast it. You then let go of the idea and let it continue to percolate in your mind. This is a process of give-and-take between the conscious and subconscious, and it takes time. Imagine you are a painter and you continue to move back and forth from your painting, placing lines, color, or images on the canvas, letting their impact settle in your mind, and then coming back to revise. As you work your ideas over the course of time, the interplay of conscious and subconscious brings greater clarity. You start to see exactly the direction you need to pursue. You have to be careful to let the process take its full course, and you have to be prepared to return to the process if your ideas, once formed, don't seem quite perfected.

It is always tempting to jump at the first idea that coalesces in your thoughts. Sometimes jumping on the idea can be part of the process, as you experiment with the idea and it leads you elsewhere. At other times, you can get fixated on an idea that you think is terrific. You will bend over backward to make the idea successful, even in the face of evidence saying that it is time to go back to the drawing board. This happens to all of us. We fall in love with our ideas. Falling in love puts a damper on the whole process of working an idea. Now we are faced with finding ways to make the idea work. The idea has gone from fluid to frozen. This is a trap into which all inventors frequently fall. The process of working an idea is not comfortable. Our natural instincts are to grab the idea, set it down, and stabilize it. We have to suspend our desire to do this for a time until we feel the idea has reached a level of maturity where it can stand above considered variations or competitors. How much time? I can't say. All I can say is that once the idea is considered to be ready and full-formed, you shouldn't fall in love with it. Always be prepared, at a moment's notice, to rework it.

I think about a glass sculptor who is constantly reworking her sculpture, forming the glass, adding color, and with every addition, needing to place the glass back in the kiln in order to make it malleable. It is similar with ideas. We need to constantly work and rework them, form and re-form them, until it is absolutely clear that we have what we want.

Action versus Thought: The Importance of Doing

Most of us cannot do everything in our heads. We can take creative thought only so far and then we need to act. The action can be as simple as writing or drawing—moving ideas from thought to paper. Or it can be building and experimentation. Often, we need to run experiments in order to validate our ideas or direction. The result of experimentation brings us back into the idea cycle with more information. Sometimes, experimentation is necessary to break a "thought loop." A thought loop is similar to an infinite loop in computing: a cycle of thoughts that keeps repeating without providing new direction or results. For most of us, imagination has its limits, and, eventually, we need to reduce ideas to practice. We can make something as simple as a rough sketch or as complex as a working prototype. This concretizes the idea and automatically shifts our brain into new considerations. How many times have you awakened from a vivid and realistic dream only to see that it could never happen in real life? Giving physical form to an idea can have the same effect. As part of your creative process, you need to test your ideas periodically in a concrete way to see if they can stand the light of day. However, even after you have built a working prototype of an idea, the design process is not finished. Even, and especially, if your prototype works beautifully, you can continue to use it as a base to imagine improvements. The processes described in this chapter should be viewed as a continuum rather than a progression. While one step might lead to the next, the next step can always lead back to the previous one.

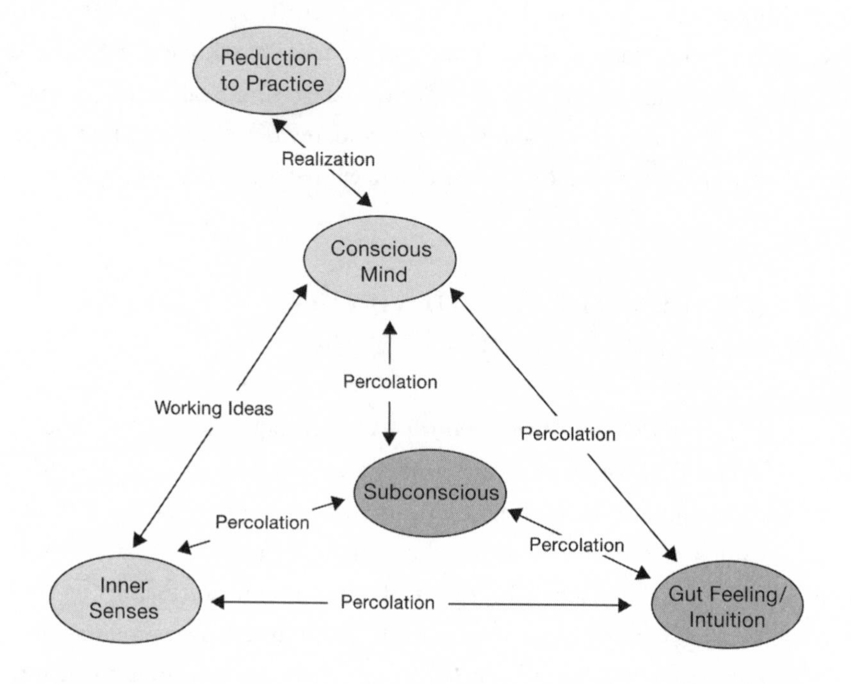

Figure 3.1: The creative process is an ongoing interaction of conscious and subconscious thought as ideas are formed. Light ovals represent conscious thought or action and dark ovals represent subconscious thought.

In this chapter, I have described various tools to access or enhance our creativity as inventors. If I had to sum up everything in a single word, that word would be *sensitivity*. In order to access these parts of our creative minds, we have to be sensitive to their existence and give them the nurturance they require. Your subconscious processing of ideas, your gut feelings, your inner senses are always there. If you cultivate and pay attention to them, they become important tools in your toolbox for generating creative ideas.

Chapter 4

THE PROCESS OF INVENTION

THE FLASH OF INSPIRATION

On my first day as a new employee at AT&T Bell Labs in the fall of 1984, I was given the assignment to develop a new computer input device to enter medical information about a patient from a hospital room. This was part of a large project to move all hospital records from paper to electronic format. Of course, computers and bar code readers existed at the time, but the challenge was developing a means of inputting patient data at the patient's bedside. Some work had already been done on this project, and the result was a somewhat clunky computer CRT (cathode-ray tube) terminal with a touch screen attached to it. This was considered state-of-the-art technology at the time. My assignment was to improve upon it.

Back then, Bell Labs was primarily the research arm of AT&T. I was given a desk, a computer console that was tied into a mainframe computer, and a large stack of documentation, piled high on my otherwise clean desk. The documentation concerned everything I would need to know about life at Bell Labs and ranged from policies and procedures to the details of UNIX command sequences. Somewhere inside the stack was a phonebook of all the people who worked at Bell Labs, with their locations and extensions. I looked at the stack of documents, then back to my clean desk, thought about my assignment for a second, and then performed my first creative act: I fished out the phonebook from the pile and looked up the one person (a colleague from graduate school) I knew in this vast organization. I found his number, called him, and arranged to meet for lunch. After all, what else was there to do? The stack of books was too intimidating; the desk was too clean; I had all my pens and pads already. I looked at my watch—only an hour and a half to go before lunch.

The problem of designing a "bedside terminal," as it was called, to enter patient information, was presented to me when I interviewed for the job. Of course, I said that I was sure I could do it, but now that I was here and looking at what they had come up with to date, I frankly wasn't sure what else to do. The CRT terminal was large, obtrusive, and heavy. The touch screen was a good idea, but where would we put this thing? Maybe it could be suspended from the ceiling?

Every time I envisioned that bulky terminal, it kept getting in the way of any kind of elegant solution. I continued to daydream, getting nowhere. My manager popped his head in, asked how I was doing, and I said, "Fine. I'm just getting my hands around the problem." But the reality was that I didn't have a clue.

I put the problem out of my mind and started thinking about my upcoming beach vacation. I had negotiated some time off before starting the job and was looking forward to my trip to the Caribbean. The time passed more quickly than expected, and soon it was time for lunch.

I had to drive to an adjacent facility to meet my friend. He was working on a computer-related project for another division. I found the building and negotiated my way through the labyrinth of hallways to his lab. Before going to lunch he volunteered to show me what he was working on.

He showed me a new computer with a flat-screen display encased in it. The computer was reputed to be more powerful than its predecessors, and the screen used electroluminescence as its light source. This was not the first time I had seen a flat screen, but this one was dazzling in its brightness and intensity.

As we walked out of the building for lunch, the design of the new bedside terminal popped into my head, almost complete. It was simple: I just needed to cut that screen off from my friend's computer, put a touch screen

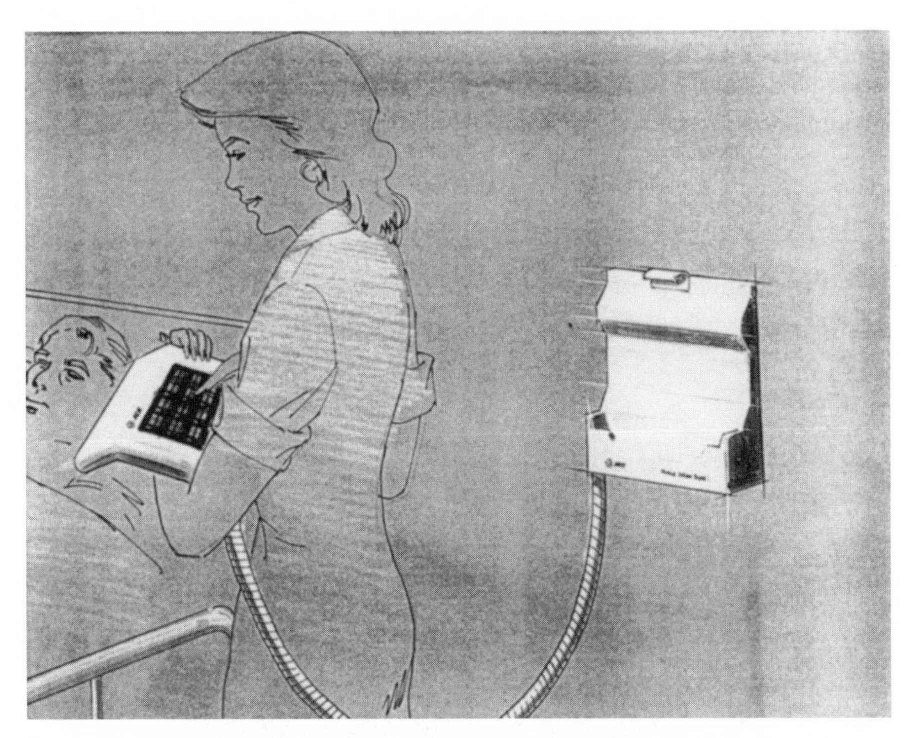

Figure 4.1a: Artist's rendering of the Patient Care Terminal.
Reprinted with permission of Alcatel-Lucent USA Inc.

on top of it, and it would be like an electronic notepad. It would have minimal electronics inside and would attach to the main computer via a coiled telephone wire in the same way a telephone handset is tethered to its base. (This was AT&T, after all.) It would be lightweight enough to be cradled in the arm of a nurse or doctor. Since most of the input information could be menu driven, the need for a keyboard would be minimal. Perhaps we could illuminate the image of a keyboard on the screen and use the touch screen to type. When not being used, this electronic notebook would hang from a hook on the wall, and the electronics to drive the terminal would be likewise mounted to the wall or placed somewhere out of the way.

This came to me in a flash. I had only been employed for several hours. This electronic notebook I envisioned became the new Bedside, or Patient Care Terminal. It was essentially completely conceptualized in a few brief minutes.

In truth, I had thought about this problem from the day I interviewed for the job. I had no great solution, but the problem was firmly implanted in my mind. The serendipity of my lunch meeting provided the spark that was needed to ignite a solution. My friend's use of the electroluminescent screen was completely different from the way I would use it. However, seeing it triggered a flash of creative thought within me, and by the time we left the building for lunch, I had essentially completed what I was hired to do.

The Process

Let's take a closer look at this process. In previous chapters we looked at the concepts of problem statements, paradigms, and the traits necessary to see what I called *the hidden obvious*. In this work assignment, the problem statement spoke of creating a "bedside terminal into which a nurse or doctor could enter patient information." The work done to date used a CRT screen combined with a touch-sensitive surface. This served to reinforce my paradigm that a data input terminal meant a cathode-ray tube with a glass screen housed in a large plastic enclosure. This kind of terminal was both bulky and heavy. Being stuck in that paradigm, one could only look for ways to make such a large object easier to negotiate around a hospital bedside—thus the idea of suspending it on a flexible mount of some kind.

I don't think I would have been able to break that paradigm had I not actually seen in front of me an example that was outside my preconception of a data entry terminal. Although I had been aware that flat screens existed, my mental focus on a CRT terminal had blocked them from consideration. Visiting my friend's lab and seeing such a terminal—even though it was embedded in a computer system in a very different way than I would imagine its use—enabled me to break out of my paradigm of a conventional CRT box.

Active awareness comes into play here as well. Although I did not enter my friend's lab in search of a solution to my problem, my mind was open to all information coming from any source that might play a role in solving my problem. This awareness was not a conscious search for a solution but more of a background, almost passive, constant observing. It was an *"I'm*

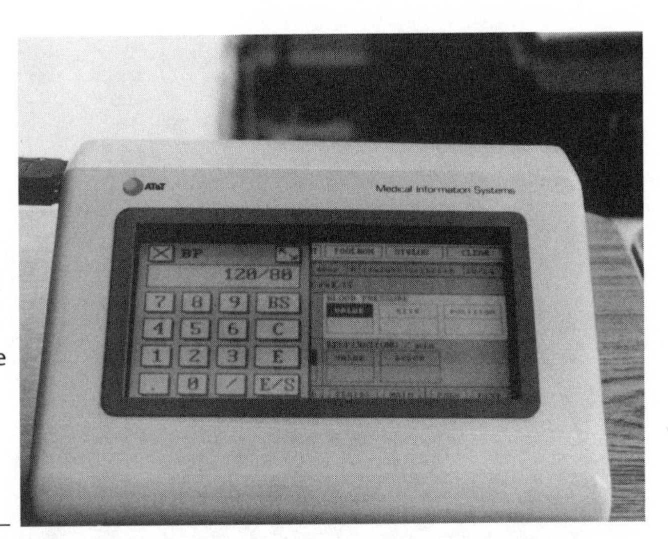

Figure 4.1b: Electroluminescent display on prototype unit. *Reprinted with permission of Alcatel-Lucent USA Inc.*

always looking and I'll know it when I see it" awareness. When I saw the screen, my interest was instantly piqued, even though I didn't know exactly how I would use it. The electroluminescent display served as the seed from which the flower—the electronic notebook—bloomed.

I was not an expert in display terminals, touch screens, or the electronics that drive them. However, I was sufficiently technically fluent to understand the scope of the problem and how to find the answers. I knew that I would have to find a way to keep the weight of this new electronic notebook to a minimum by separating the power supply and as much of the electronics as possible from the actual notebook. I understood the possibilities and limitations of what I was trying to do, and even though my expertise in the area was limited, I was able to zero in on the critical issues and research work-

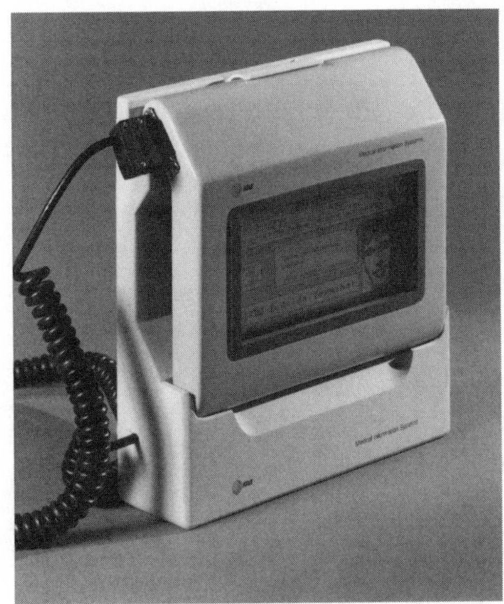

Figure 4.1c: Working prototype in mount. *Reprinted with permission of Alcatel-Lucent USA Inc.*

Figure 4.1d: The author demonstrating a mock-up of the Patient Care Terminal. *Reprinted with permission of Alcatel-Lucent USA Inc.*

able solutions. As the invention took form, it required a team of engineers to support its development.[1]

Invention doesn't always come in a single flash. But it usually comes in flashes or leaps. Whether large or small, these leaps take us to places that are off the map, out of the normal purview of connected thought. Incubation, or thinking about an idea or problem, is what makes these leaps possible, although they will not be a direct and controlled product of that thinking. My thinking about the design of the bedside terminal primed my mind for a creative burst. Even though I was fixated on using a CRT, I was subconsciously dissatisfied with that solution. In this case, my process of inventing required seeing a flat-panel display to break the mental dam and let the flow of creativity take its course.

Steve Wozniak, the cofounder of Apple Computer and the inventor of the Apple II computer, recounts a similar paradigm-shifting experience upon seeing a computer using color for the first time. He had not considered the use of color in a microcomputer as a possibility. His paradigm was that computer screens displayed information in black and white. Microcomputers, the smallest in computing power, all used black-and-white displays. He reflected that the first two small computers with color displays, his Apple II and the "Color Dazzler" from Cromenco, were inspired by the same demonstration:

> In the early days of microcomputers, the only two products to come out with color were the "Color Dazzler" from Cromenco and the Apple II. And it turns out that we both came out of the Homebrew Computer Club [an information exchange club for people interested in computers] in Palo Alto. It was a funny thing. . . . Some people from a company called Sphere brought a minicomputer [a computer larger and more powerful than a microcomputer] connected to a color TV. They showed the first color graphics that probably most of us in the room had ever seen or only imagined that a computer could produce.
>
> I just sat there thinking, I can't believe I'm seeing something like this. Probably, those people from Cromenco were sitting there just as I was: thinking what an unbelievable thing it was. For about two minutes everyone sat quietly, watching this TV draw color circles.
>
> It was just one of those things. It's hard to explain. But if that hadn't happened to me, the Apple II probably would have never had color—or even been an Apple II.[2]

I am sure that if Steve Wozniak had not been immersed in the design of new computers, he could have seen this demonstration, been momentarily impressed, and then filed it away in his brain. The fact that he was *actively aware* led him to realize the potential significance of what he saw. Seeing color displayed on a computer—a minicomputer, which was very different from the one he was designing—provided the spark that helped him form and crystallize his vision of the Apple II. Interestingly, as he points out, another computer designer in the audience had the same flash of inspiration that led to the development of a competitive product.

The Resonance of an Idea

Once we get the inspiration, what do we do with it? As discussed in chapter 2, part of the process of inventing is the courage required to create something new. This courage can be bolstered by the resonance of the idea. This is an idea, perhaps a phrase or a visual image, that continues to occupy your thoughts. Instead of diminishing with time, as many ideas do, a resonating idea continues to reverberate in your mind. In my example of the design of the Patient Care Terminal, after I had the initial inspiration, the idea continued to resonate in my mind. There were many technical hurdles to overcome, but the idea seemed so "right," that the obstacles became secondary. With time, they could be solved. My father, Edward Paley, started an entire business based on a phrase that he had read in an obscure technical magazine. The phrase was "contamination control," and in his mind, he envisioned the possibilities of a business with multiple products that he would invent using contamination control as the theme. The phrase continued to resonate. Invention after invention was inspired by the idea of contamination control as the company he founded on this theme, Texwipe, grew into a sizable corporation. An idea that resonates reinforces itself, allowing the inventor to overcome self-doubt and have the courage to plow ahead. Usually an idea or vision is momentarily clear and then gets clouded with doubt. Ideas that resonate, like materials that vibrate at a certain natural frequency, tend to rise above the doubt. They become stronger and more ingrained, and demand action.

Sources of Innovation

Where do the ideas that become new inventions begin? There are numerous sources for these ideas. We will consider five major sources of inspiration:

1. Applications of life experience,
2. Making the familiar new,

3. Applying physical principles,
4. Accidental and chance observations, and
5. Learning from nature.

Applications of Life Experience

To say that your creative ideas are a product of your life experience seems to be an obvious statement. However, invention often represents life experience and knowledge applied in a completely different context. For example, if an electrician, an orthopedic surgeon, and a graphic artist were all presented with a mechanical design problem, they would probably all solve it in a different way, influenced by their background, knowledge, and life experiences. The chances are that their solutions would be very different than that of a mechanical engineer. We bring ourselves to whatever problems we try to solve, and this gives individuality and uniqueness to the solution.

Invention often sprouts from the cracks between areas of knowledge. Typically, a person who is blessed with an instinctive curiosity has studied a variety of unrelated areas. These areas are often so disparate that they come together only through the spark of problem solving or invention. Nathan Pritikin, best known for his revolutionary work on diet and heart disease, was a self-taught engineer. Before that he ran a photography business in Chicago, taking photographs at weddings or other events, developing them, and selling the prints. His interest in engineering led him to study the subject on his own. His biographer tells the story of how he applied his knowledge of photography to a completely different problem—etching closely spaced lines in glass for the War Department during the Second World War:

> He learned that the airforce was having trouble with a certain vital part of the Nordham [*sic*] Bombsight. The part, called a reticle, was a circular piece of glass with parallel engraved lines spaced 1/1000th of an inch apart. . . . No method had been created yet to ensure accurate

> spacing between the lines at such minute distances. In the early 1940s, reticles were produced using a stylus to engrave lines into glass. The process was done mechanically, but because the lines were engraved so closely together, any vibration in the room would affect the stylus and throw off the line. This was partially compensated for by placing the engraving machine on hydraulic supports which absorbed most vibration, but little could be done to eliminate all vibration and thus there was a high degree of error in the reticles. . . . *Pritikin knew nothing about reticles or the technology that was used to produce them. For the next 12 months he studied the reticle and the technology necessary to produce it* [italics mine]. . . . He pored over books on such subjects as metal and glass etching and engraving, photoengraving, printing, lithography, intaglio, letterpress, rotogravure, electroplating, and metallic deposition. He also became an expert in photographic emulsions and the latest techniques in high-resolution photography. When he wasn't studying the material in the library, he was conducting experiments in each of these areas at his Flash Foto office on North Dearborn Street.
>
> Pritikin knew that if he were going to improve upon the existing reticle, *he would have to come up with a new way of putting an image on glass* [italics mine]. . . . By the end of 1942, he had an idea. He began by reproducing the exact image of the reticle he wanted on a large six-foot-square piece of paper. At that size he could ensure that the lines were perfectly straight and parallel. He then took a photograph of the image and reduced it to a little bigger than a postage stamp, thus bringing the lines to within 1/1000th of an inch of each other. The small image was the exact size and pattern of the reticle he wanted to put on glass. He then blackened the sheet with a light sensitive chemical called resist. Any image of a light shined upon the resist would be held, or "fixed," in the resist. Pritikin then placed the negative of his reticle over the resist and shined a light through the negative and onto the resist below. This caused the exact image of the reticle to be fixed in the resist-covered glass. He then washed away the portion of the resist not fixed by the light. What was left was a black image of the reticle formed by the resist on the glass.[3]

Once the resist was in place marking the line pattern, Pritikin could etch the actual lines into the glass using a strong acid. Then he would remove the protective resist, leaving a series of perfectly placed lines 1/1000th of an inch apart. This photolithographic technique is now commonly used to produce photomasks and reticles for semiconductor manufacturing. Had Pritikin not been a photographer, it is doubtful that the idea of using a reverse of the common photographic enlargement technique would ever have crossed his mind.

Part of the process of inventing is imbuing oneself with a wide range of areas of knowledge. This separates the inventor from the technologist who studies one particular area in-depth. The more distinct areas of knowledge the inventor can draw on, the more apt he is to use them in unexpected and inventive ways. Ted Hoff, the inventor of the microprocessor, cites a similar experience:

> Consider, for example, this analog memory cell that I came up with that made use of electroplating. It was a mixture of chemistry and electrical engineering, and it was all done for adaptive concepts that we were doing via computer simulation. It was an interesting blend of skills and technologies because I had had all this chemistry as a kid and was able to solve what was nominally an electrical engineering problem.[4]

Hoff had an uncle who was a chemist and who inspired him to study chemistry as a hobby when he was younger. He didn't pursue it professionally, but he retained a knowledge of chemistry that was unexpectedly useful in designing an analog memory cell that seemed to others to be strictly an electrical engineering problem.

We bring our past knowledge and experiences to everything we do in life. The inventor should be conscious of this and should unhesitatingly pursue all areas of curiosity, even if there doesn't seem to be an apparent immediate use for the knowledge.

Making the Familiar New

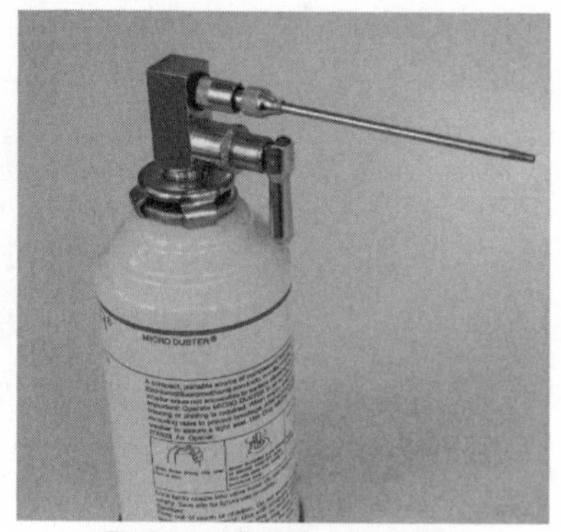

Figure 4.2: The Texwipe MicroDuster. *Photo by author.*

Inventors do not always create from scratch. Most of the time, invention involves taking something that is already known and making it new in some way. A different way of using the familiar can be as significant an invention as the original idea. The MicroDuster, mentioned in chapter 2, recast an existing nozzle system for dental irrigation as a cleaning device for critical manufacturing environments by combining it with a can of compressed gas. The novelty of the invention in this case was not the hardware but the application.

The computer is another example of making the familiar new through application. Originally, computers were used solely as scientific calculating machines. As the industry began to mature, computers were "reinvented" for use as business machines. They were then reinvented once more for use as sophisticated typewriters (word processors) and toys (video games) for the hobbyist. Today, with the advent of the Internet, computers are used by many as a primary tool for interpersonal communication and social networking, business transactions, online research, and entertainment. In each case, the familiar was reinvented to encompass a brand-new use. The flexibility to reinvent in this case is even given a name—*software.*

Figure 4.3: Apple iPhone. (2007). *Photo courtesy of Apple Inc.*

The name implies an easy malleability so that completely novel uses for the basic computer can be invented by anyone skilled in the art. One can also say that by changing the form of this invention, shrinking it down so that it can be held in the hand to produce the PalmPilot PDA, for example, the basic computer has been reinvented once

Figure 4.4: IBM 360 (1960s). *Photo courtesy of NASA.*

again. Apple's iPhone has taken this a step further by adding the ability to communicate both through phone and Internet to a handheld portable computer. This sample of computer-based inventions is by no means inclusive of all that has been done. But these devices illustrate that by finding a completely different use for an existing invention, the invention is made new again.

Applying Physical Principles

Applying known physical, chemical, or biological principles to a new use is an important method of invention. This is another way to make the familiar new. Again, a knowledge of the physical, chemical, and biological world is a necessity for the inventor so that basic principles can be used in new ways.

Xerography is an example of taking a simple principle from the physical world and using it to achieve a new and dramatic result. Xerography works on the principle that opposite charges attract each other. This principle is used several times in a complex process whereby an image is copied from one paper to another. The process begins by first transferring the image to a specially coated drum. The drum is light sensitive in that light hitting any portion of it, after it has been charged, will neutralize the charge. The drum is charged initially, and light is reflected off the image to be copied onto the drum. The drum rotation is synchronized with the illumination of the original image. Light reflecting off the white parts of the image dispels charge on the drum, and light absorbed by the black parts does not affect charge. The surface of the drum is selectively discharged in this way so that an exact charge layout of the image to be copied remains. The drum is then used to attract oppositely charged toner particles that, due to electrostatic attraction, attach only to the charged areas of the drum. The toner particles attached to the drum now form a replica of the original image. A sheet of paper is then passed between the drum and a charging corona wire. The corona wire charges the sheet of paper with a greater charge density than the drum, thereby attracting the oppositely charged toner particles to the paper. They are then fused to the paper using heat or pressure. The drum is then discharged and remaining toner cleaned off.

This process is a great example of variations on a single physical theme. We see the principle of charging and discharging materials so that oppositely charged materials will attract as the mode of operation for this invention. The physical principle is simple and widely known; the application is completely novel and is one of the most unique and significant inventions of the twentieth century.

The use of basic concepts as building blocks for inventions is a classic way to invent. These concepts, which are basically how things work, are the inventor's vocabulary. When the inventor examines a problem, she needs to have these ways of doing things at her fingertips. Of course, this is not easy. To continue with the language metaphor, most people have a limited spoken vocabulary, which is just a small fraction of the dictionary. We keep only words we use on a regular basis in the forefront of our minds. When we describe something, we immediately reach for the most commonly used pieces of language, perhaps neglecting words that are much better suited for the situation. With this understood, there have been attempts to develop processes that organize and systematize the inventor's vocabulary.

Genrich Altshuller, a Russian inventor and patent inspector, spent a lifetime studying Russian patents to look for underlying patterns of invention. He developed a system called *TRIZ* (a Russian acronym for Theory of Inventive Problem Solving). Altshuller's systematized techniques for inventing rely on basic principles. Based on his research of thousands of patents, he has culled a group of principles that can be applied to solving problems. He has also developed a methodology on how and where to use these principles. His methodology examines inherent contradictions in technical problems and solutions and works toward resolution. Lev Shulyak, a disciple of Altshuller and a *TRIZ* expert, writes the following in his introduction to Altshuller's book *40 Principles: TRIZ Keys to Technical Innovation*:

> Genrich Altshuller developed the 40 Principles more than 20 years ago. He and his team of associates reviewed thousands of worldwide patents selected specifically from leading industries for the inventive nature of their solutions. . . . Altshuller found that technical problems could be solved by utilizing principles previously used to solve similar problems in other inventive situations. For example: A "wearing problem" in the manufacture of an abrasive product, and a "wearing problem" with the cutting edge of a backhoe bucket, were both solved utilizing the principle of *segmentation*. Altshuller was able to identify 40 such principles from his analysis of successful inventions.[5]

The more you immerse yourself in the basics of physics, chemistry, and biology, the greater likelihood of your being able to draw on the most appropriate principle to solve an inventive problem. The fact that Chester Carlson was able to think about electrostatic charge as a way of transferring images—not an association most of us would make—shows the power of employing such basic principles.

Let's look at another example of extending basic principles. The density of hot air is less than that of cooler air. As a result, hot air rises. The inventor, having this basic principle somewhere in his mind as part of his inventive vocabulary, sees a problem and puts this principle to work. The attics of most houses are natural problem areas. They accumulate both heat and moisture, which can cause mold, rotting, and deforming of structural elements, in addition to making it difficult to cool the rest of the house. The sun beating down on the roof causes the attic air to heat up, and there is no way for the heat to quickly dissipate. Attic fans have been installed to mitigate this problem, but a far more elegant solution was discovered based on the principle of relative air density. The ridge and soffit vent system allows the natural process of heated air rising to enable air circulation through a continuous partly covered vent in the ridge of the roof. A series of soffit vents around the house allows cooler outside air to replace the escaping heated air, creating a continuous convective air current through the attic. This very simple application exemplifies how the inventor can look to solve problems by applying fundamental physical principles. Others have taken this solution a step further by designing special baffles for the roof vent that cause a pressure differential when wind blows across the roof. Based on Bernoulli's principle, these baffles, shaped like an airfoil, create an area of lower pressure, and this further enhances airflow out of the ridge vent. Again, the application of a basic principle of physics is used to improve the invention.

Physical principles can be extended in other ways as well. If we understand how something works and can model it, we can look to either optimize it or gain new functionality. Sometimes, an invention can be figuring out how to use built-in properties in ways not originally intended. This can best be explained through example. Parallax Corporation, a manufacturer of educational electronics, produces an educational robotics kit called the BOE-BOT. In this kit they include a Panasonic IR sensor, which senses the presence of modulated infrared light. This is similar to the sensors used in televisions that receive information from remote-control units. The sensor provided by Parallax, called the PNA 4602M, senses the presence or absence of infrared light modulated at a frequency of 38.5 kHz. The sensor,

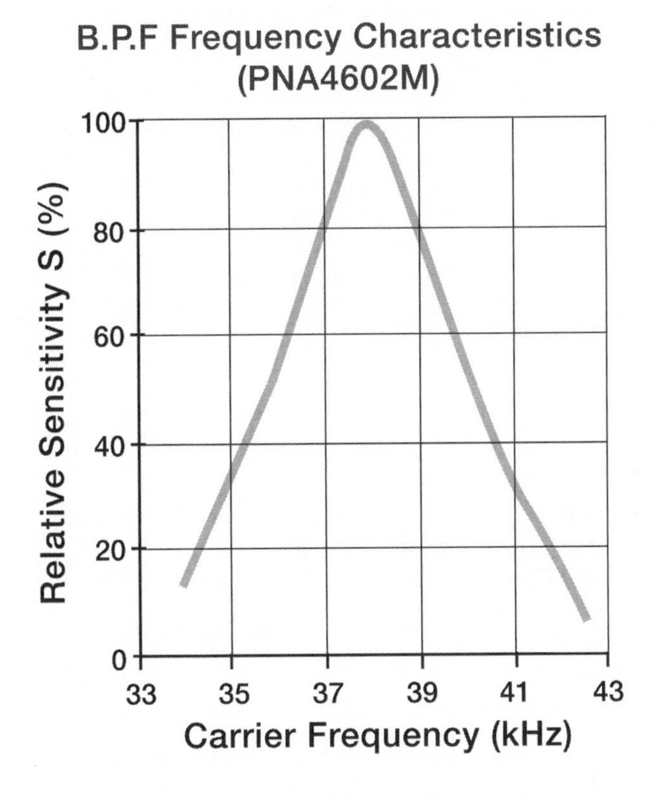

Figure 4.5: Frequency sensitivity for the PNA 4602M infrared sensor.
Reproduced by permission of Panasonic Corporation of North America.

used in combination with an infrared transmitter, is used in experiments for obstacle avoidance (detecting IR-transmitted light bouncing off an obstacle in the robot's path), edge detection, and so on.

It turns out that the PNA 4602M, a relatively inexpensive sensor, has a broader frequency response than might be desired. This kind of sensor is supposed to filter out all infrared frequencies except 38.5 kHz. However, this sensor is not a very precise filter. The frequency response graph in figure 4.5 shows that the PNA 4602M can respond to frequencies from a little over 33 kHz to almost 43 kHz. This makes the sensor less than ideal at filtering out extraneous IR frequencies.

What does this limitation have to do with invention?

Sometimes invention involves taking advantage of physical properties that are considered deficiencies. That is exactly what the people at Parallax did with the frequency response of the PNA 4602M sensor.

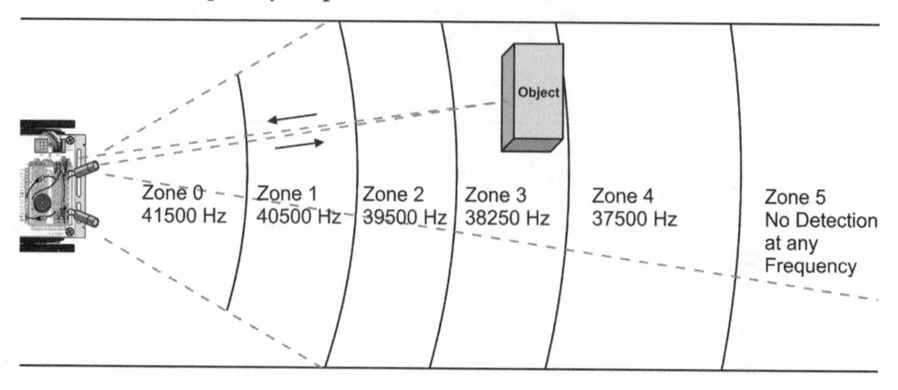

Figure 4.6: Robot distance measurement based on frequency.
Courtesy of Parallax Inc.

As mentioned previously, this sensor is used in an educational robotics kit in combination with an IR transmitter. Normally, the transmitter would be programmed to transmit at the peak detection frequency of 38.5 kHz. The transmitted signal will bounce off any object in the path of the robot and return to the receiver, where it is detected. This way, the robot will be able to detect objects in its path. If we instead transmit at a frequency slightly less or greater than the peak frequency, the sensor will be less sensitive to the transmitted signal. This affects how far away we can detect the transmitted signal. An object that would be detected a certain distance away with the transmission set at 38.5 kHz would only be detected at a closer distance if the transmission is off-peak. The engineers saw something positive and novel that they could do with this anomaly: they created a distance sensor. They reasoned that by sending out multiple frequencies from the transmitter corresponding to various points on the sensitivity curve, this inexpensive sensor could be used to detect not only the presence of an obstacle but also the robot's distance from that obstacle. The less-sensitive frequencies would produce a response only when the object was close, and the more sensitive frequencies would produce a response when the object was both close or far away. By summing the responses over a variety of frequencies, distance from the object could be determined.

This clever and inventive use of a basic property of this device, normally seen as a limitation, is an example of how understanding the basic behavior of a system can lead to creative and novel uses.

ACCIDENTAL AND CHANCE OBSERVATIONS

"In the field of observation, chance favors only the prepared mind."
—Louis Pasteur[6]

Sometimes the familiar becomes new purely by accident. Invention, in this case, is the ability to recognize the accident as something significant. One of the most famous accident stories is the discovery of penicillin by Alexander Fleming in 1928. Fleming was studying the disease-causing *Staphylococcus* bacteria. He had been frustrated by the tendency of mold contamination to ruin his bacterial cultures. On one occasion, he noticed that the bacteria growing near the mold on a contaminated culture plate had broken down. Fleming suspected that the mold was producing a substance that was toxic to the bacteria. He was able to extract the antibacterial substance from the mold and named it penicillin after its source, the common bread mold *Penicillium*.

Fleming's discovery came from outside contamination to his experiments. He could not keep the air in his laboratory sterile and hence was constantly throwing away experiments due to mold contamination. His recognition of the strange reaction that one of his culture dishes had to a particular mold contamination led to the isolation and discovery of penicillin and a future Nobel Prize. While the discovery of penicillin can be looked upon as a chance event, had Fleming not spent years researching the properties of *Staphylococcus* bacteria, he would almost certainly not have recognized this mishap for what it was. His active awareness and his immersion in the field of bacteriology combined to allow him to see something that might have normally just slipped away.

Accidental discovery, as Louis Pasteur said, "favors the prepared mind."

When we do something and the unexpected happens, our tendency is to ignore it and continue to pursue our initial train of thought or action. Recognizing the productive accident or side effect requires tremendous insight and willingness to change direction or paradigm. The story of the invention of the microwave oven shows how large the paradigm shift can be.

Dr. Percy Spencer was a radar researcher with Raytheon Corporation engaged in developing new radar technology for the military. No doubt his paradigm for the use of radar revolved around detecting airplanes in flight. Raytheon is a prime military contractor and is one of many companies depended upon by the Department of Defense for new technologies. A chance event, the melting of a candy bar, could simply have gone unnoticed in the larger effort to achieve better radar. But this accident was noticed and led to a new industry.

> Like many of today's great inventions, the microwave oven was a by-product of another technology. It was during a radar-related research project around 1946 that Dr. Percy Spencer, a self-taught engineer with the Raytheon Corporation, noticed something very unusual. He was testing a new vacuum tube called a magnetron when he discovered that the candy bar in his pocket had melted. This intrigued Dr. Spencer, so he tried another experiment. This time he placed some popcorn kernels near the tube and, perhaps standing a little farther away, he watched with an inventive sparkle in his eye as the popcorn sputtered, cracked and popped all over his lab.
>
> The next morning, Scientist Spencer decided to put the magnetron tube near an egg. Spencer was joined by a curious colleague, and they both watched as the egg began to tremor and quake. The rapid temperature rise within the egg was causing tremendous internal pressure. Evidently the curious colleague moved in for a closer look just as the egg exploded and splattered hot yolk all over his amazed face. The face of Spencer lit up with a logical scientific conclusion: the melted candy bar, the popcorn, and now the exploding egg, were all attributable to exposure to low-density microwave energy. Thus, if an egg can be cooked that quickly, why not other foods? Experimentation began . . .
>
> Dr. Spencer fashioned a metal box with an opening into which he fed microwave power. The energy entering the box was unable to

> escape, thereby creating a higher density electromagnetic field. When food was placed in the box and microwave energy fed in, the temperature of the food rose very rapidly. Dr. Spencer had invented what was to revolutionize cooking, and form the basis of a multimillion dollar industry, the microwave oven.[7]

Recognizing an accident is just the beginning. The next thing is to convince yourself and others that the accident has significance. Fleming pursued his ideas for over ten years with minimal success. His tests were inconclusive, and it was not until 1940 that a second team of independent researchers was able to concentrate penicillin sufficiently to show consistent benefit. Likewise, using microwaves to heat food was an idea that took years to become accepted. The inventor is often the one who launches the ship but is not necessarily the one who captains it to its destination. This is especially true with accidental discoveries whose nature is more tenuous than intentional discoveries. The inventor or discoverer often finds herself on strange footing in a new and unfamiliar area. The invention is often taken to fruition by others who, although they did not make the discovery, are better positioned to act upon it.

In addition to physical inventions, significant accidental discoveries are often made in the marketplace. Invention can make the familiar new by the discovery of a new marketplace for an existing product. The example of the invention of chewing gum, described in chapter 1, is essentially the discovery of a marketplace for an existing substance. While the sap from chicle trees never did work out as a substitute for rubber for the tire industry, the accidental discovery of its use as chewing gum opened up a whole new area of confection.

I experienced an accidental discovery of a marketplace while working at the Texwipe Company. Among the products we produced were foil-enclosed towelettes that were saturated with 91 percent isopropyl alcohol. These were used in the computer and electronics industry for removing residual oils and other contaminants. One day we received a letter from a police officer in Florida who had somehow obtained a sample of this product. He wanted to know how he could get more. He went on to explain in the letter:

"Every day I deal with criminals and the scum of the earth. . . . When I come home to my family at night, I need to *disinfect* myself from my work."

The key word in the letter is "disinfect." We previously served only the computer industry, but this letter opened up the possibility of another marketplace in our minds. Since we had the means of producing presaturated towelettes, we could easily develop a line of disinfecting towelettes. Perhaps it was the earthy humor of the letter that got our attention long enough to do a double-take and think about the possibility. We certainly received many letters about our products during the course of business, and most did not spawn new markets for us. But this discovery of an "accidental side use" of this product led our company into an entirely new marketplace.

Accidents happen all the time; most can be safely ignored. However, the occasional accident can lead to invention—sometimes a tremendous breakthrough. The challenge is recognizing the significance of the accidental side effect even if it is not in the area you are currently pursuing. The difficulty is taking that step back, appreciating the event for what it can be, and changing direction, if necessary, to take action.

LEARNING FROM NATURE

With all the inventive genius that humankind can muster, we can never create anything as remarkable as the natural world. What we can do is borrow from the natural world as well as learn from it. As inventors, nature can serve us in many ways: we can directly use elements of nature—for example, making medicines directly from plants; we can imitate nature—as George de Mestral was inspired to do when inventing Velcro; or we can simply learn from observing our natural surroundings and apply that learning to an area of invention. Nature is a wonderful example of almost all that we would want to invent. We just need to be observant enough, sensitive enough, and smart enough to distill what is readily before us. For instance, all our efforts in developing computers are paltry before the capabilities of the human brain. How does the brain work? How does it do so much with what seems to be so little? We are just at the early stages of being able to answer these questions.

Our immense efforts in computing have merely given us powerful calculating machines. We have nothing that compares to the innate processing abilities of the human brain. The animal kingdom as a whole presents a myriad of opportunities for inventors to learn. Why, for instance, are there no animals that use wheels for locomotion? What can we learn from this? What can we learn from their versatility of movement? The study of birds in flight has been a fascination of humankind since early times. How do they do it?

Only in the last 150 years have we discovered the principles behind flight. Wilbur Wright, through careful observation of birds in flight, discovered that the key to their turning was not steering via a boatlike rudder mechanism but twisting the ends of their wings in opposite directions—a discovery put to use in the Wright brothers' aircraft and in all airplanes today as ailerons.[8] The answers to many of our basic questions are openly visible in nature; we just need the clarity of vision and the powers of discriminating observation to be able to see them. The human body is a remarkable work of engineering. It is much more advanced than anything we can conceive of in the laboratory or in industry. What can we learn from it? What can we imitate? For instance, the body heals itself when damaged. Can we design machines to do the same? What about energy efficiency? Are our cars as efficient as our bodies? Why not? How do ecosystems use feedback to adapt to change? How do rivers and forests restore themselves? Is there something we can learn from this?

As you can see, there is no end to this train of thought. Nature is an encyclopedia of knowledge and examples for the inventor. Let's look at a few illustrations of inventions taken from nature.

Using Nature Directly

If we start with the idea of taking something directly from nature and using it as part of an invention, we could start with two of the oldest uses of nature. The first would be the use of fire. Fire was not invented by humans, but once discovered, its utility was clear. It served early humans as a source of heat, enabled the invention of cooking, and allowed people to see in the dark. Today we have heaters, stoves, and flashlights, among other derivative inventions. The basic invention came directly from nature and was then refined and shaped by humans to provide for specific needs. This is a common thread with inventions that are taken directly from nature.

A second ancient invention taken from nature is the harnessing of power. Originally, humans generated their own power through the use of their bodies. However, observation of nature inspired the use of animals as a power source. Not only did they take the burden off people, but they were stronger and could provide greater power for longer periods. This discovery led again to the use of animal power as a key part of many inventions, from the plow to the war chariot. Along with increased power, this discovery provided increased mobility, which had profound effects on the making of civilization. No one invented the horse, but someone invented horse travel.

Medicines are another form of invention taken directly from nature. Until the industrial age, all medicines were based on plants, insects, animals, or other life-forms with effectiveness determined by trial and error. Many of these discoveries remain with us today, such as digitalis from the foxglove plant and codeine from the opium poppy. The modern pharmaceutical industry routinely looks for new drugs from natural sources. Pacilitaxel, better known as Taxol, was discovered through a National Cancer Institute grant to the Department of Agriculture to collect samples of thousands of plants to be tested for anticancer properties.[9] The invention here is twofold: First is the discovery of application. What plant, animal, or other organism can be used in curing which disease? Second, once something from nature is identified for its curative properties, how is it extracted, concentrated, purified, and even synthesized? Remember the story of penicillin. It took ten years for chemists to figure out how to concentrate it sufficiently so that it could be used as a therapeutic agent. The pharmaceutical industry today is well aware that many of its future breakthroughs will be based on substances found in nature, and it invests commensurately in searching them out.

Wood, clay, and stones have long been the basic building blocks for invention. Since the dawn of civilization, we have culled from these natural materials all that we needed to create the artifacts necessary for survival and industry.

Another interesting aspect of directly using nature to assist in invention is the training and use of animals or other living organisms with special skills or qualities. These living creatures become either the invention or an integral part of it. Examples would be bacteria that turn color when exposed to pollution and bomb-detecting dogs. These are inventions that directly rely on the unique abilities found in these species. The process of invention in these (and many other similar) cases is to identify special abilities in certain species and match them with problems for which they offer an ideal and unique solution. Sometimes animals or other living organisms can be used as part of a larger system in which they are combined with other technologies. The famous behavioral psychologist B. F. Skinner initiated a project during the Second World War called Project Pigeon, code-named Project Orcon (for "organic control"). This project predated reliable guidance systems for bombs, and Skinner looked to nature to invent such a system. The idea of this top-secret project was to train pigeons to peck at a target through a lens in the nosecone of a missile or bomb. Their pecking would act as input to an electronic feedback system that would direct the bomb to the target.

> Started during World War II, Project Orcon (for organic control) was a try-anything approach to solving some then-current problems. Guidance systems for homing missiles were being easily countermeasured and the Navy thought animals might have potential as a jam-proof control element. Pigeons were selected for trial because they were light, easily obtainable and adaptable. Their job was to ride inside a missile and peck at an image of a target picked up by a lens in the missile's nose. The pigeon's pecking of the target image was translated into an error signal that corrected the simulated missile's simulated flight.

Trainee pigeons were started out in the primary trainer pecking at slowly moving targets. They were rewarded with corn for each hit and quickly learned that good pecking meant more food. Eventually pigeons were able to track a target jumping back and forth at five inches per second for 80 seconds, without a break. Peck frequency turned out to be four per second, and more than 80 percent of the pecks were within a quarter inch of the target. The training conditions simulated missile-flight speeds of about 400 miles per hour.

Figure 4.7: A "trainee" pigeon for Project Orcon. *Photo from iStockphoto.*

The image was shown under a glass screen coated with stannic oxide to make it electrically conducting. The target was moved by a small mirror controlled by a servo. The control circuits were such that if the pigeon stopped tracking, the target image would drift rapidly away from the center of the screen. This forced the pigeon to correct not only his own pecking errors, but those introduced by the yawing of the missile. It turned out that 55.3 percent of the runs made were successful—that is, the pigeons were able to keep the target image on their screens for the duration of more than half their flights.

If pigeon guidance did not get very far in the Navy, it did have one valuable offshoot. The electrically conducting glass was later used in many radar displays.[10]

It's comical to read about such an approach today, but the idea of looking to harness unique capabilities found in species throughout the natural world is important for inventors to consider.

Observation and Imitation

Using something directly from the natural world to solve a problem is an approach that has been used for centuries and continues to be used today with even greater degrees of sophistication. However, since the advent of the industrial age, we now have another way of using nature—that is to observe and imitate it. We can observe how natural processes work and then imitate them in the products and processes we design. The story of Wilbur Wright observing bird flight is a perfect example of this. He did not use actual birds in the airplanes he designed with his brother Orville, but he learned how they turned in the sky by studying how they changed the position of their wingtips, and he applied that knowledge to his design. Nature is a great teacher, and what we observe has been honed to great purpose and efficiency. It is for us to comprehend working principles from nature and to apply them to our inventions.

The invention of Velcro shows how careful observation and imitation led to a breakthrough invention. Plant burrs are specifically designed to attach to things temporarily so that they will be transported and eventually released in order to spread the seed of the plant. It only took the right inquisitive person to notice that this same idea could be applied to a fastener.

The study of biological processes as models for invention can give tremendous insight. While we tend to think mechanistically, biological processes are not limited in that way. Many scientists and inventors look to these processes as something to be imitated.

When Percy Shaw invented the first embedded road reflector, he realized that it would need to be self-cleaning. Being embedded in the pavement of a dirty roadway would cause it to quickly get caked with grime and lose its utility. Having a maintenance crew go out and clean each reflector would be prohibitively costly and impractical. He looked to the biology of the human eye and its ability to continuously clean itself as a solution that he could mimic for his road reflector.

> He therefore took further inspiration from its living analogue: by mimicking the actions of both eyelids and tear glands. He mounted four of the glass spheres, two pairs facing in opposite directions, in a raised latex moulding which in turn was set into a cast iron "shoe." This was then asphalted into the centre of the road. The master stroke was in the

Figure 4.8: Catseye road reflector.
Photo by ELIOT2000.

way these devices cleaned themselves. The shoe was designed to hold rainwater, and the moulding had a lip positioned just below the two glass spheres: the main part of the housing could be deformed to move past it. When a car drove over the stud, it depressed the housing into the shoe, causing the lenses to wipe on the lip. It also compressed the rainwater held in the shoe below the moulding and caused it to squirt upwards, cleaning the spheres as it did so.[11]

Another interesting aspect of this invention is its name: Catseye. The inventor is said to have derived both the idea and the name from the experience of driving his car on a foggy road one night. He suddenly saw his headlights reflected back to him in the eyes of a cat. He quickly swerved to avoid the cat, and in doing so, also avoided going off the side of the road. He thought about the natural phenomenon of light reflected back from a cat's eye and applied this idea to his invention.

Materials scientists often look to biology for clues on how to develop new materials. While our understanding of materials science seems advanced and sophisticated, new materials can be invented by imitating natural processes that are far superior to those that we have today. Scientists and engineers studying ceramics have always been plagued by a paradox: the harder they make a ceramic, the more brittle it becomes. They have attacked the brittleness problem by making the material grains smaller. This has helped, but it has not solved the problem. Ceramics have a multitude of uses such as knife blades, electrical insulators, bearings, heat-resistant coatings, and artificial bone. They are known for their hardness and ability to withstand harsh environments. Research into better ceramics is a key area in the field of materials science, and solving the problem of fracture due to brittleness is of primary importance.

There are materials in nature that are as hard as synthetic ceramics, but they are not as brittle. Researchers have started to look to biology in order to understand these materials and to see whether they can mimic them. Janine Benyus describes some of this research in great detail in her book *Biomimicry*.

> But how does nature create that microstructure? And how can we do the same? Answering those questions is at the very heart of what biomimics are trying to do. "We want to do more than just copy down the angles and architecture of nature's designs or build our materials in their image," says ceramist Paul Calvert from the University of Arizona Materials Laboratory in Tucson. "What we really want to do is imitate the manufacturing process, that is, how organisms manage to grow, for instance, perfect crystals and form them into structures that work."
>
> . . . A few years ago, Calvert thought it was time to energize imaginations, so he and other biomimics began to look at natural designs. They discovered plenty of examples of biological organisms that, like the abalone, sport hard body parts made from a mixture of inorganic minerals and organic polymers. . . . In all these cases, nature's crystals are finer, more densely packed, more intricately structured and better suited to their tasks than our ceramics or metals are suited to ours.[12]

Besides studying abalone, the search for hard materials that are also lightweight and resilient has led scientists to look at other creatures in an attempt to develop new paradigms. Herb Waite of the University of California at Santa Barbara describes his study of the clam worm, whose jaws are both hard and wear-resistant yet contain no mineral content—something completely contrary to our understanding of how to make hard materials.

> He [Waite] explained that it is usually axiomatic that if (in materials science) you want something hard, stiff and wear-resistant, it needs to have mineral in it—minerals are the best materials for making something stiff. In nature's best cutting structure, vertebrate tooth enamel, the mineral content is 95 percent. Yet the clamworms do not use any mineral for hardening, thus revealing a new paradigm in nature and sparking new ideas for materials science.

"The jaws are much lighter than if they were calcified," said Waite. "If you want something lightweight then you reduce reliance on minerals. I'm not sure why, in their watery environment, the clamworms' jaws need to be light." Yet the jaws are lightweight, wear-resistant biomaterials without mineral.

"It's not that technology will copy verbatim what the worm does," said Waite. "But we are interested in knowing concepts of how organics interface with transition metals to produce functional material."[13]

Ultimately these studies and others like them will give technologists the information necessary to mimic natural processes to produce the next generation of super-hard, lightweight, and resilient materials.

Adhesion is another area in which scientists can learn by observing nature. For example, the effort to make a glue that will adhere underwater or in moist environments has long eluded scientists. In her book, Benyus quotes Herb Waite's description of how this has already been invented.

> Nature invents and we invent. In fact, I think that humans and all other life forms have been evolving toward similar points, but other organisms are simply further along than we are. They have already faced and solved the problems that we are grappling with. For instance, *edulis* (a species of mussel), wanting to eat in a tidal zone, had to manufacture a glue that could stick to anything underwater. We know how tough that is, because our adhesive industry has been struggling for years to come up with an adhesive that can work in moist environments and stick to anything. It's still out of reach. Mussels are light-years ahead of us.[14]

Waite then proceeds to discuss his research into mussel adhesion. The elegance of the solutions to the various problems of adhesion is a wonder to perceive—from surface preparation to attachment to eventual release and biodegradability of the glue. For example, the mussel secretes a foam as part of its attachment mechanism. The foam gives flexibility as well as protection from crack propagation. This allows the mussel to adhere to surfaces that contract and expand with tidal movement. The mussel creates the foam without the use of a gaseous blowing agent. This technique has only recently been understood and utilized by industry with the need to eliminate CFCs from the production of commercial foams. This simple organism has within it the answers to a whole variety of problems that challenge modern industry.

Recently, the chemical composition of the mussel adhesive has been identified and spliced into the genes of the common soybean to make an inexpensive, water-resistant wood glue. The discovery, made by Professor Kaichang Li of Oregon State University, is described below:

> "Soy beans, from which tofu are [*sic*] made, are a crop that's abundantly produced in the U.S. and has a very high content of protein," Li said. Soy protein is inexpensive and renewable, but it lacks the unique amino acid with phenolic hydroxyl groups that provide adhesive properties. Li's research group was able to add these amino acids to soy protein, and make it work like a mussel-protein adhesive. Then they began to develop other strong and water-resistant wood adhesives from renewable natural materials using mussel protein as a model.
>
> The new wood adhesives are made from natural resources such as soy flour and lignin. They may replace the formaldehyde-based wood adhesives currently used to make some wood composite products such as plywood, oriented strand board, particle board, and laminated veneer lumber products—all major components of home construction and many other uses.[15]

Scientists have turned to nature to discover how new and better materials can be invented. Nature's solutions are often in plain sight, and it is up to the ingenuity of scientists to figure out how they work. Imitating biology is a powerful tool for invention. It takes a careful observer and a mind open to the unique and clever ways that nature operates.

Imitating biology is not only a microscopic endeavor. Animal behavior can also provide the key to invention. The design of mobile intelligent robots is an immense challenge and another example of how we can learn by imitating nature. To design a robot to perform a task in a set environment is not difficult. The problem is clearly defined and always the same;

an example of this would be a robotic welder on an automobile assembly line. But what if we wanted to design a robot that had to perform a task in a multitude of environments—places so varied that the designer could never anticipate all of them? Say, for example, the robot's task was to pick up a pencil and place it on a desk. If the scenario was always the same—the same pencil, the same desk, the same physical proximity between the two—this could be a straightforward task. But what if the environment could not be anticipated? What if the robot was to be sold worldwide, and its job was to perform this task wherever pencils and desks were found? The programming challenge would be incredibly complex.

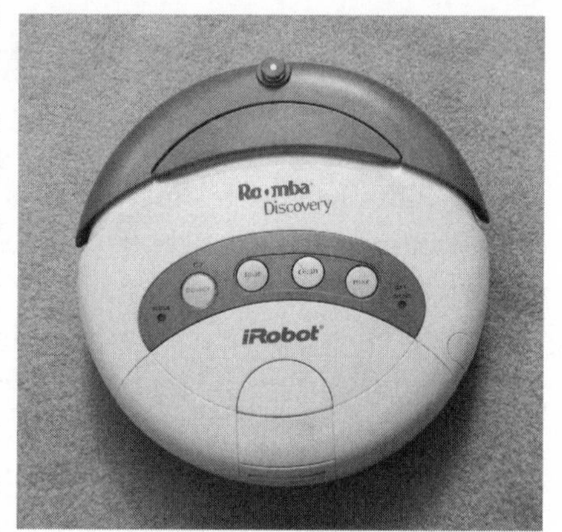

Figure 4.9: The Roomba robotic vacuum cleaner. *Photo courtesy of Shawnc.*

This is the type of challenge faced by the designers of the Roomba robotic vacuum cleaner. Its task was to vacuum the floors in a house—not just one specific layout, floor surface, and furniture arrangement, but an almost infinite variety of surfaces. How could the designers give this robot the intelligence to perform its task effectively, given the untold number of situations in which it had to work? The designers turned to nature for the answer.

Joseph Jones, one of the inventors of the Roomba, writes in his book *Robot Programming, A Practical Guide to Behavior-Based Robotics*, that the answer came from studying insects. These are creatures with extremely small brains, yet they manage to perform their tasks in a wide variety of situations and environments. Jones writes:

> Insects fascinated the mobile robot folks. They noted that these creatures are a marvel in a miniscule package. In a complex and dangerous world, insects manage to find food, shelter and mates. Insects escape from predators. Insects navigate their world and don't get lost. And sometimes insects even seem to cooperate in building large structures

> and in performing other impressive feats. Yet insects have the tiniest of brains. For many insects, sight is accomplished using primitive vision systems—systems that boast fewer pixels than cheap video cameras. What were dumb bugs doing that put our best robots to shame?[16]

The study of insects enabled the development of behavior-based robotics—a methodology for creating software that gives the robot the flexibility to adapt to differing environments and challenges. This system, like that observed with insects, sacrifices efficiency for flexibility; specificity for robustness. In the end, behavior-based programming was the breakthrough that not only led to the commercially successful Roomba vacuum but was also instrumental in the development of the Mars Sojourner Rover, which explored the unknown surface of Mars.

We instinctively tend to look at processes as mechanical and serial—a single cause producing an effect, whereas natural processes often act in parallel with multiple causes producing multiple simultaneous effects, which then produce secondary effects. This creates systems that are in constant dynamic equilibrium. These systems often use many simple behaviors to achieve complex results. Natural systems are robust, self-healing, and failure-tolerant. Think of a vine growing by a fence. If you cut it, it sprouts new shoots; if its growth is blocked by an obstacle, it changes its path; it uses the fence as an aid to growth, but it is not completely dependent on it. This basic plant has no intelligence, and yet it is supremely adaptable.

The transition from the nineteenth century to the twenty-first century is one of moving from a mechanistic way of thinking to a more organic way of thinking. Whether through direct use or imitation, nature provides us with a treasure trove of solutions just waiting to be applied.

The Processes of Invention

The elements of invention come from a wide variety of sources. The inventor's process is to expand his view to encompass as many sources as possible and yet be able to grasp, at the right "aha" moment, the crucial idea and use it to form his invention. We have seen that invention can come from random accidents or carefully applied basic physical principles; from the workings of nature or new applications for an existing manufactured device. Invention can be a sudden realization or the result of methodical study. The process of invention is really the *processes* of invention. The common thread is that inventors constantly have their eyes open. They are always on the lookout. They don't always know exactly what they are looking for, but they know it when they find it.

PART II

DESIGN AND INVENTION

Chapter 5

SIMPLICITY

Design is the embodiment of invention. Whether the invention is a paper clip or a computer chip, its design is the incarnation of the creative ideas that spawned it. Design is the process of making something hard and definite emerge from the swirling mists of creativity. If design embodies invention, then *simplicity*, *elegance*, and *robustness* are the goals of good design. I introduced these principles in the first chapter and will now revisit them in greater depth. I find that these principles, as applied to invention, often take the form of questions rather than statements: "Does my design reflect with absolute simplicity the essence of my idea?" "Have I solved the problem with the utmost elegance?" "Is my solution robust enough to withstand the unanticipated?"

Let's explore these principles in detail and see how they apply to invention.

PROFOUND SIMPLICITY

As the world we live in becomes increasingly complex, a great deal is being written about the virtues of simplicity. Today, many corporations throughout the world are adopting simplicity as their guiding corporate mantra. We have all experienced the frustration of too much complexity: confusing road signs, new products overloaded with features, baffling government forms, or the ubiquitous remote controls for entertainment systems. Complexity was once thought to add value to a product or system by providing more options or information. However, the human mind is unable to keep too much information in the forefront, and so clarity of purpose is often sacrificed. The ability to accomplish simple tasks is lost in a sea of possibilities.

Simplicity in invention and design, however, has a somewhat different meaning. I see simplicity as getting to the essence of something. Simplicity starts with the problem statement. In chapter 1, I spoke about how we must first examine a problem and make sure it is the one we need to solve, and not one that is merely peripheral to our central concern.

We often need to examine a problem and dissect it into its components, then reconstruct it in its simplest manifestation. We need to ask ourselves: "Have I expressed the essence of the problem that I need to solve? Have I pared it down to its core?" Once the problem is stated in an essential and unequivocal way, the solution that follows can mirror the problem in directness and simplicity.

Not every problem is simple, nor is every solution. Recall Einstein's admonition:

> Everything should be made as simple as possible, *but not simpler*.

We should strive to simplify as much as we can without compromising what needs to be accomplished. Simple does not mean dumb. Simple means direct and efficient—with nothing extraneous. Simplicity is the graceful movements of a professional athlete that look so easy, so natural, and yet took years of training and practice. Simplicity is an endpoint. It is also a beginning.

In his book *Profound Simplicity*, psychologist Dr. Will Schutz describes the journey from simplicity to simplicity. He says this journey necessarily travels through the realms of complexity before returning to a fundamental or profound simplicity. He writes:

> Understanding evolves through three phases: simplistic, complex, and profoundly simple.[1]

He speaks of discovery as moving from the simple to the complex to the profoundly simple.

> A powerful example of profound simplicity—the one that awakened me to the concept—is in the 1912 work of Bertrand Russell and Alfred North Whitehead, *Principia Mathematica*. In that book, the authors considered the enormous complexity of mathematics and demonstrated that it could be reduced to five simple, logical operations. From these observations, all mathematics could be derived.[2]

This idea had a great effect on Schutz. The idea that complexity could be reduced to all-encompassing axioms meant that many seemingly diffuse problems could ultimately be tied together with core solutions—solutions that addressed the essence of the problem.

A scientific researcher I knew discussed the search for profound simplicity as it applied to his research: "We start off with the simple hypothesis, our experimentation leads to more and more complex hypotheses, and in the end, we make our discovery, which is profoundly simple."

The inventor, too, will follow this same path when searching for solutions. He often starts with a simple idea and then, in pursuit of the optimal solution, adds layers of complexity. Still not satisfied, he continues to add complexity in an attempt to get at the solution. Finally, a profoundly simple solution presents itself (often due to his subconscious ruminations), and this is the essence of his invention.

"Is my solution the simplest it can be?" The inventor needs to ask this question over and over as he gives form to his ideas.

Let's look at a few examples of profound simplicity.

The Paper Clip

The paper clip, as noted in chapter 1, seems to be the epitome of simplicity in invention. It is essentially a piece of bent wire that uses the torsional properties inherent in the wire to act as a clamp. The simplicity of the device is that it uses basic properties of its material of construction to perform its task. Its construction material, a small piece of metal wire, does not suggest the function of clamping. It is the profoundly simple invention of a few bends in the wire that gives it this novel function. Would any of us have come up with this startlingly simple invention if presented with the need to temporarily hold sheets of paper together? Probably not. The simplicity of using the innate torsion of a wire to apply clamping pressure becomes obvious only once it is seen. Interestingly, this torsional principle is now commonly used in diverse applications, from car suspensions to CD players. The paper clip is a wonder of simplicity in invention and a model for all of us to emulate in our own inventions.

The Laser Printer

Sometimes an inventor has an idea, and as he proceeds to give it form through design, he moves from the simplicity of the idea into realms of complexity that continue to multiply and confound. Gary Starkweather, the inventor of the laser printer, found himself in such a situation. His idea seemed simple. He felt that he could make a radical improvement to xerography by using a laser rather than ordinary white light to illuminate the image of the page to be copied onto the xerographic drum. The original image would not have to come from a sheet of paper but could come in the form of electronic data from a computer. Starkweather built such a device, but he soon realized that the precision needed for the alignment of the optics would make his invention extremely complex and exorbitantly expensive. As he grappled with the increasing complexity, he became frustrated and discouraged about his progress. Weeks turned into months, and still he could not get past this problem. Finally, one day, with nothing better to do, he went back to basics—to simple things he had learned in his first year of the study of optics. From the simple came the breakthrough that led him to success. Michael Hiltzik describes this breakthrough in his book *Dealers of Lightning*:

> Still, the most troublesome problem . . . fell squarely within the domain of traditional optics. Starkweather knew that if the mirrored facets were even microscopically out of alignment, the scan lines would be out of place and the resulting image distorted or unintelligible. . . . In visual terms, the mirrors could not be off by more than the diameter of a dime as viewed from a mile away.

For more than two months he wrestled with the puzzle. . . . One day he was sitting glumly in his optical lab. The walls were painted black and the lights were dimmed in deference to a photoreceptor drum mounted nearby, as sensitive to overexposure as a photographic plate. Starkweather doodled on a pad, revisiting the rudimentary principles of optics he had learned as a first-year student at Michigan State. What was the conventional means of refracting light? The prism, of course. He sketched out a pyramid of prisms, one on top of another, each one smaller than the one below to accommodate the sharper angle of necessary deflection. He held the page at arm's length and realized the prisms reminded him of something out of the old textbooks: an ordinary cylindrical lens, wide in the middle and narrowed at top and bottom. "I remember saying to myself, 'Be careful, this may not work. It's way too easy.' I showed it to one of my lab assistants and he said, '*Isn't that a little too simple?* [italics mine]'"

It *was* simple. But it was also dazzlingly effective. Starkweather's brainstorm was that a cylindrical lens interposed at a proper distance between the disk and the photoreceptor drum would catch a beam coming in too high or low and deflect it back to the proper point on the drum, exactly as an eyeglass lens refocuses the image of a landscape onto a person's misaligned retina.[3]

As complex as the laser printer is—even more complex, I am certain, in its first iteration—going from the land of complexity back to the simple proved to be the key to making this device a commercially viable technology. Simplicity is perhaps most important in complex devices that rely on tight tolerances where failure lurks at every turn. When we discuss robustness, we will see the importance that simplicity plays in making a design reliable.

The Ridge and Soffit Vent System

Reliability and simplicity are often joined at the hip. As with the paper clip, simplicity can provide a robust solution that relies on natural properties. As inventors, we should always look first to natural properties to see how they can be used to our advantage. An excellent example of incorporating naturally occurring properties is the aforementioned ridge and soffit vent system for cooling house attics. This system works simply by the natural convection caused by heated air rising. There are no moving parts, nothing to break down (perhaps the soffit vents can become clogged over time, but they can easily be inspected and cleaned). The solution is an exemplar of simplicity. The solution it replaces—large electric attic fans—is more complex and costly and is subject to failure. The simple solution, which uses properties of the natural environment to its advantage, shows how simplicity makes designs more reliable. Living in the twenty-first century, our instinct is to gravitate toward technology to solve problems. If presented with the problem of cooling an attic, we would tend to create a technological solution requiring fans or blowers and, of course, electricity. The simple—more elegant and robust—solution requires no new technology. It requires only that we "go with the flow."

The Sealed-Edge Wiper

Sometimes simplicity is the act of breaking a complex operation or design into its simpler components. I dealt with such a situation while working for the Texwipe Company, where I was involved in the invention of a new wiping material that would be used to remove contamination during the manufacture of computer chips. The wiping material was made from polyester fiber that was knitted into a fabric and then cut to size. The problem with this process, however, was that simply cutting the material left the raw edges exposed, which then led to the shedding of deleterious fibers and particles. Our solution was to partially melt the edges of the wiping material so that no shedding would occur. We decided that a modified version of ultrasonic welding, a technology used to fuse plastics together, would be the right approach to sealing the edges. The system we were developing involved three untested technologies: (1) a method to precisely control a flexible and unstable web of material, (2) an ultrasonic sealing technique using a patterned drum that would seal all four edges as it rotated, and (3) a cutting technique done on the same steel drum that would cut along the melted edges and separate the finished product from the web. Needless to say, with all the new technologies coming together, the system was at a high risk for failure. We had assumed that the failures would not be fundamental but incremental. That is, any failures that might occur could be overcome through engineering. We were wrong. While the complexities of web handling and cutting fell into the category of engineering problems, the ultrasonic welding of the edges of the material did not. This manner of welding had not been attempted before, but it should have worked—at least in theory. We examined the problem from all angles and collaborated with the supplier of the ultrasonic equipment without success. While much was understood about the ultrasonic welding of thicker plastics, the fusing of a thin, knitted material such as ours was more of an art than a science.

Our project was definitely at the periphery. We had already preannounced the product with a delivery date of six months into the future. The reception, based on handmade prototypes, was extremely positive, and demand was growing. The only problem was that we could not mass-produce the product. Two months before the release date, we received a call from IBM. The company had just sampled a prototype of our product and found that it solved a major manufacturing problem. IBM then proceeded to place the largest order ever received in Texwipe's history. Like any company, we were loath to turn down an order, especially one of that magnitude from such a prestigious corporation.

Needless to say, the pressure was on. We realized that we had pursued all our technology options with no success. We could not get the ultrasonics to work as they needed to on our drum. With time running out, we needed a solution, and that solution was to simplify. We knew that ultrasonics could effectively seal the material along a single axis. We had also solved the various issues with controlling the unwieldy web and cutting the material. Our solution to the ultrasonic problem was to break a single complex process into two completely separate but much simpler processes, each sealing one set of parallel edges at a time. Although this was less efficient from a labor and manufacturing standpoint, it worked! For all our initial attempts at developing the ultimate, most efficient method to manufacture the product, a simplified—albeit less elegant and more labor-intensive—approach turned out to be the best way to mass-produce the product. Now we could reliably begin production and, through engineering, continue to refine the process.

Often, combining operations results in simplicity. In Texwipe's case, simplicity required that we separate a complex operation into its simpler component parts.

The iPod and the Walkman

Apple's iPod is often cited as a model for simplicity of design in an age when the complexity of consumer electronics is mind-numbing. The iPod's design is no accident but rather is the result of the careful study of how people interact with machines—nowadays called "interaction design." The iPod was designed for intuitive use; it was also designed to be operated without intensive study and while pursuing another activity, such as exercise. The design is simple, intuitive, and focused. Using a circular motion of the thumb or finger to scroll and a single press of the push-button to select is so simple and instinctive that most people learn to use the device quickly. The simplicity is repeated as you work your way through levels of choices. Scroll and select. Scroll and select. Only once you select a song and it starts playing is complexity added by the options available with multiple presses of the select or arrow buttons. You can always return to a previous level by pressing the menu button, which will take you back from whence you came. The simplicity of function is mirrored in the design, where nothing is superfluous or confusing. This design paradigm has been refined and carried into even more complex products such as Apple's iPhone and its recently released iPad.

Now compare the simplicity of navigating around any of these newer products such as the iPhone to a typical cell phone of the previous generation. While most cell phones prior to the iPhone were designed to do many things, it was difficult to do anything other than make a phone call without consulting the manual. Most of these devices were examples of needless complexity. They lacked the intuitive simplicity of the iPod and its progeny.

In his book *The Laws of Simplicity*, MIT professor John Maeda describes the evolution in design of the iPod's dial. The diagram below shows three successive iterations of the dial design:

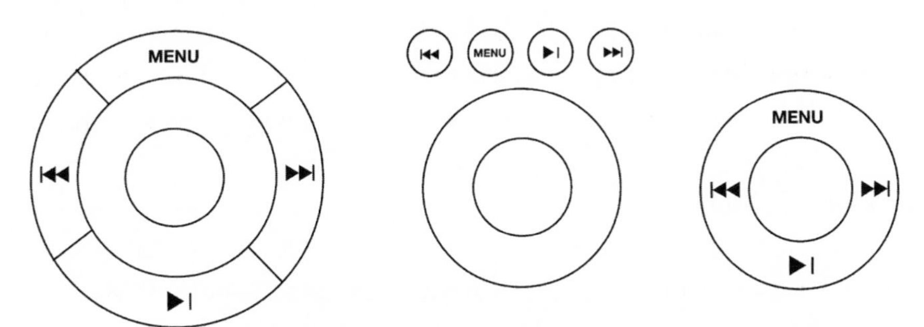

Figure 5.1: The evolution of the iPod controls.
Reprinted with permission of the MIT Press.

Maeda describes the sequence of iterations as "starting simple, then getting complex, and ending up as simple as possible."[4] This process is reminiscent of the journey from simplicity to complexity to profound simplicity described by Will Schutz earlier in this chapter.

The iPod's simplicity and ease of use stand out in the sea of miniaturized portable consumer electronic devices. The designers of the iPod homed in on the principle of simplicity to make their product a model of how easily humans can interact with computerized accessories.

The original personal music system provides another example of the principle of simplicity. The Sony Walkman, launched in 1979, created a new category of entertainment systems. Before that, stereo systems were either stationary or could be carried around, albeit somewhat awkwardly. No one had produced a system that was "wearable." The story goes that Akio Morita, the head of Sony, was sitting in his office when an engineer came in with a portable stereo tape recorder and a pair of large headphones. Morita felt that the product was too heavy and bulky and told the engineer to *simplify* it by removing the recording circuit and speaker and adding very lightweight headphones.[5] Simplifying the device seemingly reduced its functionality, but it actually changed the device into something very different—a completely new device that people could wear to listen to music.

There were many skeptics, both at Sony and among the general public, when the Walkman was announced. The feeling was that nobody would purchase a cassette player that would not record. Morita fought the skeptics, and the Walkman was released into the marketplace with tremendous success. Sales eventually reached hundreds of millions of units. In this case, simplifying led to an entirely new industry of wearable entertainment systems.

Amazon's 1-Click

The Internet is a relatively new phenomenon, and the area of Web site design is constantly evolving. One of the oldest and most successful Web sites is Amazon.com.

Amazon originated as a bookseller and has expanded to become a distributor of numerous other products. The beauty of Amazon is the simplicity of its interface. Navigating the site and purchasing a product are both extremely simple. Amazon's designers surely had the principle of simplicity in mind when creating the site. This principle seems to be the guiding force behind how easily and directly shoppers can perform the basic functions of browsing among the products, finding the one they wish to purchase, and making the purchase. I can imagine the development of Amazon's site: With each iteration of design, the question is asked: "How can I make this process simpler?" In their quest for simplicity, the designers at Amazon came up with their patented "1-click system." In retrospect, 1-click seems like an obvious concept; however, at the time, all Internet retailers used the virtual shopping cart system for online purchases. Perhaps they felt that storing proprietary information such as a credit card number online would scare off their customers. In its quest to simplify the path from selection of a product to its purchase, Amazon realized that the simplest way would be to just "click" on the product. The thinking behind 1-click was captured in an interview with Jeff Bezos, the CEO of Amazon: "At the time he [Bezos] came up with 1-click shopping, everyone was locked into the mindset of the shopping cart metaphor. On the Web, he realized, all you had to do was point and click on an item, and it was yours."[6]

I'm sure the simplicity of 1-click has resulted in millions of dollars in sales for Amazon that might not have happened if only the shopping cart was used. The shopping cart is more complex and requires time, typing, and thought. Industry studies show that between 60 and 65 percent of online shopping baskets are abandoned before they are checked out.[7]

The genius of inventive thought here is something so simple as to be obvious. If it is easier to make purchases, people will buy more. I often think of this while waiting in line at retail stores. Making customers wait in line discourages them from buying. Retailers complain about the Internet hurting their businesses, but they don't seem to be looking at ways they can encourage more purchasing by simplifying the buying process.

THE SIMPLE AND THE COMPLEX

The beauty of simplicity can only be reflected in the mirror of complexity. In fact, the two need to coexist. John Maeda, in his book *The Laws of Simplicity*, describes this paradox:

> *Simplicity and Complexity need each other.* The more complexity there is in the market, the more something simpler stands out. And because technology will only continue to grow in complexity, there is a clear economic benefit to adopting a strategy of simplicity that will help set your product apart. That said, establishing a feeling of simplicity in design requires making complexity consciously available in some form. The relationship can be manifest in either the same object or experience, or in contrast with other offerings in the same category—like the simplicity of the iPod in comparison to its more complex competitors in the MP3 player market.[8]

The brilliance of simplicity is measured against complexity. There are many ways of doing something, but when a simple solution is found—one that mitigates the implied complexity of the task—it stands out. The complexity of how something could have been done is there in the background for all to imagine.

Some machines or devices are necessarily complex—an airplane or an automobile or a computer, for example. Referring back to the quote by Albert Einstein, "Make everything as simple as possible, but not simpler," we can see that these systems, even in their ultimate complexity, have been designed to be as simple as possible. And if one looks at where the principle of simplicity is most clearly applied, it is in the human interface to these complex machines. Here simplicity needs to master complexity. The controls of an automobile are such that they require no knowledge of the complexity of the machine. It is not necessary to understand the nuances of computerized fuel injection systems to drive a car. The complexity of the machine is controlled in the simplest way.

Imagine that the complexity of an automobile is reflected in its control systems. Let's say that the dashboard is filled with computer screens with readouts from every microcontroller in the car (there are twenty to more than sixty microcomputers in most modern cars), and that part of driving would involve constantly scanning the screens in order to monitor the status of all onboard automotive systems. Driving would assume a new level of complexity, and not many individuals would qualify for driver's licenses. Of course, this is an unnecessary addition of complexity. The systems can monitor themselves and alert the driver in case of failure. I am continuously amazed that the evolution of the automobile has led to a machine—with all its power and potential destructive capacity—that the general population, regardless of education and with minimal training, can operate in a safe and reliable manner.

Any complex system that requires human operation must be clear and unambiguous. Simplicity in this regard doesn't necessarily mean minimalism. Consider a typical digital wristwatch. My son has one, and every so often he asks me to reset it. The watch has a few buttons on the side and a large LCD dial. For the life of me, I cannot reset the watch without referring to the operations manual. The three unlabeled buttons on the watch control many different functions. One has to operate them in just the right sequence and manner to address any one of the functions. One can marvel at the simplicity of having so much controlled by so little, but this type of minimalism becomes a parody of simplicity. There is no way to figure out how to operate the watch without reading the detailed and complex instructions.

Figure 5.2: The control room at the Three Mile Island nuclear power plant.
Reprinted with permission of AP/Wide World Photos.

The other extreme is a device with too many controls, especially controls that look the same but do different things. Here we have ambiguity and complexity. The control room at the infamous Three Mile Island nuclear plant is an example of this. In an emergency, the operators could not effectively use the controls to contain the escalating problem.

Simplicity in the design of things with which people interact is predicated on clarity and intuitiveness. Controls should be intuitive to use, and the effects of their use should be immediately clear to the operator. If possible, one or more of the senses should be engaged via feedback from the control. For instance, when you type on a keyboard, how long must you press down a key? How hard must you press it? When electronic keyboards were first created for computers, they made no sound, as there was technically no reason for sound. In fact, the sound of hitting keys was considered to be a throwback to the days of mechanical typewriters. However, as users, we need the aural feedback of the click of the keys when we type. This way we know that we have positively hit the key. Therefore, manufacturers soon added an electronic click to the keyboard as an auditory aid to typing. Most keyboards today have tactile feedback as well. There is an initial mechanical resistance that must be overcome when pressing a key. Once this happens, the key travels to its stopping place with minimal resistance. Overcoming this initial resistance lets us know that the key is pressed.

Fly-by-wire controls in airplanes are another example of designed-in feedback to promote clarity. These controls are purposely designed to give the feel of mechanical controls—even though they are completely electronic. This gives the pilot the necessary kinesthetic feedback to control the airplane more effectively. The list goes on and on. A great deal has been learned and written about human-factors engineering and interaction design. I have provided only a few illustrative examples. When creating something that will be operated by a person, simplicity promotes clarity and ease of use. A manufacturing engineer I knew once summed it up by saying, "Make it easier to do it right than to do it wrong."

Zero Mass Design

Simplicity as a modus vivendi for invention can be measured in terms of "bang for the buck." The "bang" in this case is functionality. By functionality, I mean how many different things we can do with the invention. The number of features added year after year to common software programs doubtlessly increases their functionality. But they become so complex that users must spend much time and effort to take advantage of all the features. We know that we can get a great deal of functionality by increasing the complexity of a design, but what can we get from something simple? Our paper clip is simple and, as previously mentioned, has tremendous functionality. The idea that functionality can be gained through both complexity *and* simplicity has been put forth by Dr. David Thornburg in his manuscript *Zero Mass Design*.[9] Dr. Thornburg describes the dynamic of gaining functionality as a U-shaped curve where functionality can be achieved from increased complexity or extreme simplicity. He considers functionality gained from extreme simplicity to be *inventive design*, and functionality gained through increased complexity to be *engineering design*. Figure 5.3 illustrates this concept.

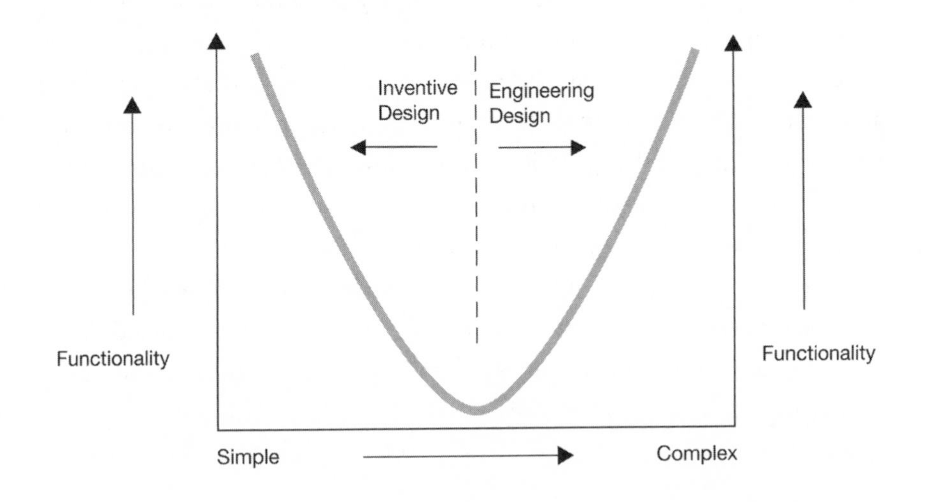

Figure 5.3: Functionality can increase through complexity or simplicity. *Graph courtesy of David Thornburg.*

In traditional design, as one increases complexity, functionality also increases. However, in zero mass design, functionality can increase as one simplifies. A paper clip is an example of zero mass design. It is very simple, and yet it has functionality that extends well beyond its intended use.

Dr. Thornburg describes the principle of zero mass design: "The principle of Zero Mass Design is that one starts with the simplest design possible, even if it fails to work. The final design will be an evolutionary refinement from that point."[10] The idea of simplicity increasing functionality is a profound one. Inventors and designers often find that once they capture the absolute essence of what they are trying to accomplish, their creation is not only simple but has functionality that exceeds their original aim.

Think of the Sony Walkman, described earlier. Akio Morita told his engineering group to reduce the complexity of the design by removing what were considered to be key features. However, to the surprise of many, this actually increased the functionality of the design by allowing people to wear the cassette player, thus creating an entirely new product.

The interlacing of threads known as weaving is an ancient invention. It is also simple and can be done by hand or by machine. From this simple invention comes cloth, another simple invention, with uses that are too numerous to list. One of the advantages of simplicity is that the invention can be molded by the users to their needs. Functionality becomes the property of the collective imaginations of all those who use the simple invention. Clothing was probably the first obvious use for cloth (hence the name), but padding, decoration, cleaning, filtering, polishing, holding, containing, transporting, and so on were quickly added as people realized that this invention could be used to satisfy a myriad of different needs.

The idea that simplicity can increase functionality leads to the idea of *elegance*, the topic of the next chapter.

Chapter 6

ELEGANCE

THE MANY FACETS OF ELEGANCE

The word *elegant* is often defined as "stylish and graceful." We think of this meaning of elegance in terms of design, fashion, or architecture. We imagine elegance as something that is so "right," it brings its surroundings to a new state of heightened grace. The Alamillo Bridge in Spain is an example of the classic definition of elegance.

Figure 6.1: The Alamillo Bridge in Seville, Spain, designed by Santiago Calatrava. *Photo from iStockphoto.*

I am going to discuss another way of describing elegance as it applies to invention. I look at elegance as *doing the most with the least*. Elegance goes hand in hand with simplicity. An elegant invention is one that exploits simplicity to gain abundance, and this can be done in several ways. In the first part of this chapter, we will focus on elegance through ingenious simplicity. An invention can be simple but not necessarily elegant. However, most elegant inventions provide a profoundly simple yet very effective solution to a problem. Often the ingenuity of these inventions involves breaking through an existing paradigm and seeing the problem in a new way.

A second way of looking at *doing the most with the least* is to focus on the "most" part of the statement. Some inventions are so fundamental that they can be used in a multitude of different ways—often ways that were never imagined by the inventor or discoverer. I describe these as *core* inventions, and they often serve as the building blocks for future inventions. Many of these inventions are also ingeniously simple. Their elegance is shown by how many uses they have beyond their creator's original intent.

A third type of invention that *does the most with the least* is one that can modify its form, content, or behavior in response to its outside environment. These are inventions that do the "most" by changing themselves or their actions based on how they are used. Inventions of this sort fall into two broad categories: those that react or change behavior in a predictable way, and those that change in a manner impossible to predict. Predictable inventions change by adapting themselves in a foreseeable way to external conditions. Think of a spring applying a gradually greater opposing force as it is stretched. Unpredictable inventions actually grow and become something new as they are used. A prime example is the evolving World Wide Web with its proliferation of new sites.

In this chapter we will examine the many forms of elegance through example. We will start with inventions that highlight ingeniously simple solutions to problems, then we will move to fundamental or core inventions that have multiple and flexible uses. Next we will shift gears a little and look at inventions that have the ability to automatically regulate their responses based on their innate properties. Finally, we will examine inventions that can change and actually grow during their use. Each category of invention represents a different facet of elegance.

Ingenious Simplicity: The Most from the Least

What is ingenious simplicity?

In a nutshell, it is figuring out how to get the most from the least. It is simplifying a complex way of doing something and at the same time doing it better. In the previous chapter, we examined David Thornburg's concept of zero mass design, which states that functionality can actually increase with simplicity. The key to increasing functionality through ingenious simplicity is to make use of everything within the locus of the invention—the problem, the properties of materials, the outside environment, the limitations—with an eye for novel ways of using any of these to achieve your goal.

This idea is best shown through example. The inventions described below are each elegant in that they illustrate ingeniously simple solutions to complex problems. An ingeniously simple solution is often hidden right within the problem itself. The inventor must ask: "How can I make the problem work for me?"

Let the Problem Be the Solution

The ridge and soffit vent system uses the problem of hot air rising and getting trapped as the engine for its elegant solution. It uses nothing more than natural convection to gain its functionality.

A minor problem with the Panasonic PNA 4602M infrared sensor (discussed in chapter 4) became the source of an ingenious and elegant solution for measuring distance, something that the sensor is not designed to do. Again, the inventor greatly enhanced functionality by figuring out how to take advantage of a limitation with the frequency filtering of this device. The route to elegance can come by creatively turning a constraint into a solution.

A New Way to Use Old Materials

The integrated circuit uses a material—silicon—in a novel, nonobvious way to simplify and miniaturize electronic circuitry. Prior to the invention of the integrated circuit, all electronic components that were used to create a circuit were completely separate entities. Resistors, for example, were manufactured in one factory out of one set of materials, capacitors in another factory out of different materials, and the newly discovered transistors in a third factory with completely different techniques and materials. The components were then assembled to form a functional circuit in yet another process whereby they were placed on a circuit board and wired together. The engineers could see that as circuits became more complex, the required amount of space and material would grow to the point where manufacture, use, and maintenance would become impractical. If the electronics industry were to continue to grow, it needed a way to miniaturize components.

In 1958, Jack Kilby, a new engineer at Texas Instruments, was assigned to a project to make circuits more compact. Kilby started work just as the plant was shutting down for a mass employee vacation, and because he had not accumulated any vacation time, he was left alone in an almost empty building to ponder the miniaturization of circuits. He knew Texas Instruments did one thing well: the manufacture of semiconductors, which serve as the heart of a transistor. Transistors were produced using a slab of silicon or germanium as the base substrate, and Kilby wondered whether it might be possible to make the other components out of that same slab of material. The idea that resistors or capacitors could be made from semiconductor material at first seemed far-fetched, but as Kilby examined the idea, he felt that it was possible. By August 1958, he had successfully fabricated individual resistors and capacitors out of silicon and demonstrated these discrete components in a functioning circuit. The following month, he produced a working prototype of the first complete electronic circuit created from a single piece of germanium. Because all the components were built on one substrate, this invention became known as the integrated circuit.

Kilby describes his thought process in the article "What If He Had Gone on Vacation":

Figure 6.2: Jack Kilby's 1958 integrated circuit. *Photo courtesy of Texas Instruments.*

> In my discouraged mood, I began to feel that the only thing a semiconductor house could make in a cost-effective way was a semiconductor. Further thought led me to the conclusion that semiconductors were all that were really required—that resistors and capacitors, in particular, could be made from the same material as the active devices. I also realized that, since all the components could be made of a single material, they could also be made in situ, interconnected to form a complete circuit. I then quickly sketched a proposed design for a flip-flop using these components. Resistors were provided by bulk effect in the silicon, and capacitors by p-n junctions. I [then] built up a circuit using discrete silicon elements. Packaged grown-junction transistors were used. Resistors were formed by cutting small bars of silicon and etching to value. Capacitors were cut from diffused silicon power transistor wafers, metalized on both sides. . . . Although this test showed that circuits could be built with all semiconductor elements, it was not integrated. I immediately attempted to build an integrated structure as initially planned. I obtained several wafers, diffused and with contacts in place. By choosing the circuit, I was able to lay out two structures that would use the existing contacts on the wafers. The first circuit attempted was a phase-shift oscillator, a favorite demonstration vehicle for linear circuits at that time.[1]

By manipulating the material in different ways, capacitors, resistors, and transistors could all be formed from a single semiconductor substrate. The paradigm breakthrough that many different parts—previously constructed using different materials—could now be made with one material earned Kilby the Nobel Prize in Physics in 2000.

So we see that ingenious simplicity is a way of making the most from the least. Kilby's idea that everything needed for an electronic circuit could be created from a slab of silicon or germanium exemplifies elegant invention.

A Movable House?

What if you were given the problem of inventing a portable house? The house has to be relatively easy to transport but also comfortable to live in. The house should offer protection from weather, provide a means of heating and cooling, have sufficient ventilation for cooking, be easy to maintain and repair, and should be constructed from readily available materials.

Think of how you would solve this problem today. Maybe you would design a geodesic dome with segmented carbon structural elements hinged to fold up with a breathable membrane material on each triangle. What about heating, cooling, and cooking? Impressive designs have come out in recent years for lightweight tents for campers and backpackers. New materials and designs provide ease of assembly and disassembly for transport, along with comfort and requisite shelter from the elements. But these designs are derivative of a truly elegant solution to the problem: the tepee devised by the American Indians who roamed the Great Plains. The tepee—a conical tent—is an ingeniously simple way of solving the problem stated above. It elegantly uses the technology of the day to address all the housing issues faced by nomadic tribes. The tepee is made from readily available materials—wood and animal skins or fabric; it is easily assembled, disassembled, and transported; it protects its occupants

Figure 6.3: Plains Indian tepee. *Photo from iStockphoto.*

from the elements; its shape keeps it stable in harsh winds and provides a large ratio of living space to material used for construction. It has natural convective cooling (similar to our ridge and soffit vents or solar chimney), and it allows for cooking within its structure and, as smoke rises, provides adequate ventilation.

My guess is that nobody invented the tepee in one day. It probably evolved over time. However, its elegance is shown in its simplicity and efficiency. It does a lot with just a little. It makes ingenious use of readily available materials in a way that solves a multifaceted and complex problem.

Limitations Spark Ingenuity

Sometimes a problem comes wrapped in such severe constraints that elegance through ingenious simplicity becomes a necessity. The Conix One inhaler, developed by Cambridge Consultants (shown in figure 6.4), is designed to dispense dry, powdered medication in a single dose of aerosol inhalant. This product is intended to compete with the syringe as a means for administering vaccinations through inhalation rather than an injection; therefore, the inventors needed to design a device that could be manufactured as inexpensively as a syringe. Because of this significant cost constraint, the device had to be extremely simple and easy to manufacture. The energy source for dispensing the medication would have to be the recipient's intake of breath.

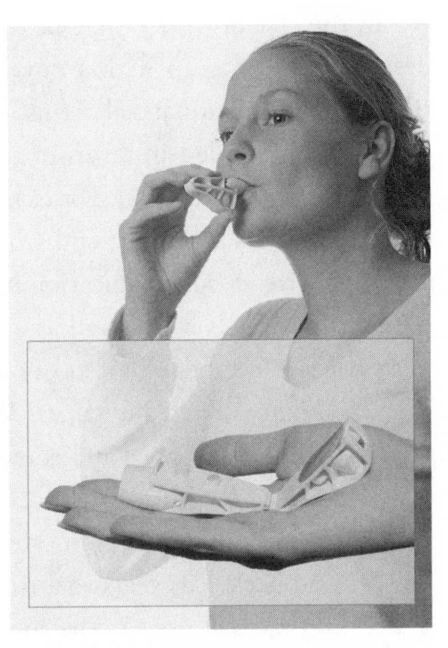

Figure 6.4: The Conix One inhaler designed by Cambridge Consultants. *Reprinted with permission of Cambridge Consultants.*

In light of these constraints, the design that resulted is a small, simple molded plastic device that is expected to cost as little as four cents to manufacture. The designers were able to develop this unique and elegant device by engineering the airflow pattern to create what they refer to as a "reverse flow cyclone," the effect of which is to deliver significantly more medication than was possible with any other existing breath-powered inhaler. The inhaler effectively separates large particles from small ones and then directs only the small ones—which can penetrate deepest—into the lung. It has no moving parts and yet delivers superior performance compared to much larger and more complex devices.[2] The designer describes it as being "akin to providing the performance of a sports car at a cost comparable with a moped."[3] The elegance of this design lies in the fact that, despite extremely harsh cost constraints, the inhaler achieves so much with so little.

In the case of the Conix One inhaler, severe cost constraints forced the designers to come up with a novel and elegant product. But cost can be thought of in nonfinancial terms, as well. In the area of communication, cost can be regarded in terms of the number of basic elements needed to communicate information. For example, the English language has twenty-six letters in its alphabet, and from those letters a vibrant and growing means of written communication has developed. The system of letters and numerical symbols works well as a means of writing, and there is no reason to reduce the number of letters or numerical symbols. But what if we didn't have the luxury of pen and paper? What if we needed to communicate language with all its nuance and complexity in a much more restricted way? We might need to radically reduce our "cost" in terms of the number of elements that we use to communicate.

Let's go back 150 years to a time when electricity was first being harnessed for practical use. At the time, it was thought that it might be possible to transmit information over electric wires. But how? Unlike the use of pen and paper, the symbols for letters could not be formed from electricity and sent over a wire. Perhaps one solution would be to make a system of twenty-six wires, each representing a separate letter. Then you simply place a voltage on the wire representing the desired letter to be transmitted. The first telegraphs, developed in England, used a system that dealt with this limitation in a similar but better way. Instead of twenty-six wires, there were five wires with a device at each end that could decode the combinations of which wires were "on" or "off" into letters of the alphabet. The encoding device would use an electromagnetic system to position a series of needles that would represent a specific letter. When a switch was pressed, the needles on the receiving device would pick up the signal and move to corresponding positions. Words could be transmitted in this manner one letter at a time. This was a natural way to think about transmitting information, but it was slow and cumbersome.

Samuel Morse, while working on the same idea of transmitting information over electric wires in the United States, developed an extremely elegant solution—a solution requiring only a single wire. His solution was to simplify the alphabet by encoding the letters into a combination of two states—a short pulse and a longer pulse—known as the "dot" and the "dash." Using sequences of dots and dashes, he developed a code for every letter and number. The advantage of this system is that it required only two elements to communicate the entirety of the language. A skilled operator could rapidly key the information. Morse's telegraph was originally designed to print on paper tape; however, the operators found that they could easily recognize and translate the inadvertent mechanical clicks the device made each time it printed a dot or a dash, thereby eliminating the need for the paper tape.

Figure 6.5: Early telegraph key.
Photo from iStockphoto.

The elegance of Morse's solution is highlighted by the limitations he was working under. Electricity had recently become usable for commercial and practical purposes. A signal on a wire could either be the presence or absence of an electric current. The paradigm of communication is that we use words, which consist of letters, and so the natural inclination would be

International Morse Code

1. A dash is equal to three dots.
2. The space between parts of the same letter is equal to one dot.
3. The space between two letters is equal to three dots.
4. The space between two words is equal to seven dots.

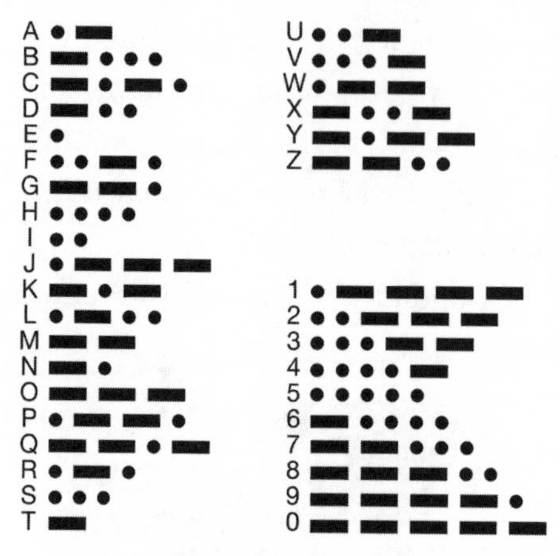

Figure 6.6: The Morse code, consisting of dots and dashes, was originally developed by Samuel Morse and Alfred Vail.

to make a device whose basic units were letters. Morse and his colleague Alfred Vail broke with the natural paradigm and instead incorporated into their design what seemed like a major limitation. If an electric wire can have only two states—on or off—then why not find a way to make these states work as the basic communication elements? The dot, a short "on" then "off," and the dash, a longer "on" then "off," formed the basis for Morse code.

Morse code was a precursor to the twentieth-century idea of binary encoding of information. The idea that information can be encoded into two basic states is the foundation of modern computing, information storage, and communications. Today, most forms of information are converted into a series of ones and zeros as the information is stored, manipulated, and transported via modern electronics. Through the Internet, this further refinement of the work of Morse and Vail has spawned a revolution in how the world communicates.

The elegance of Morse's encoding idea is based on finding an ingenious solution within the problem itself. The problem was that electricity could only be either on or off. There was no way to spell electric letters comparable to those written on the paper. Morse and Vail realized that instead of finding a complex way of trying to get around this limitation, they could use the limitation to their advantage and develop a simplified way of describing the alphabet using only the limited capability of turning the electric current on or off. Just as the Parallax engineers discovered an advantage in the limitations of the Panasonic PNA 4602M sensor, Morse and Vail developed an elegant solution based on the constraints put before them by the new technology of electric circuits.

Simple Repeated

Let's look at another example that by definition should be elegant because it is designed to do much with little. An *algorithm* is a formula or series of steps that can be used repeatedly to perform a larger goal. Typically, we see algorithms employed in computer programs.

A well-designed algorithm is elegant because with a few basic and simple steps it achieves great functionality. Imagine walking through a complex maze with high walls that you cannot see over. Here is a simple algorithm that will enable you to navigate any maze, no matter what the design. Step 1: Walk until you touch a wall of the maze. Step 2: Walk forward, keeping your hand on the wall. Continue this process of forward motion while maintaining hand contact with the wall through all its twists and turns, and you will successfully navigate the maze. Assuming the maze has an entrance and exit, this algorithm will always work. The fact that a simple algorithm will work in diverse situations gives it power and elegance. Long division uses a basic algorithm of repeated arithmetic operations to gain a complex result. The elegance of the algorithm is that it works with any pair of numbers. Algorithms are often the heart of computer programs. When one admires the elegance of a program, it is usually the algorithms that are being extolled. A well-thought-out algorithm uses simplicity to create abundance.

Each of the inventions we've examined hinges on a discovery that led to an ingeniously simple solution. The ingenious part of the simplicity is how it can achieve greater functionality than a more complicated solution. Ingenuity can come in many different ways. It can involve turning a problem or limitation into a solution, as in the ridge and soffit vents, the PNA 4602M distance measurement, or the invention of Morse code. Or it can come from reaching into the basic properties of materials and finding a new way of using them, as was the case with using silicon or germanium for the integrated circuit. Sometimes profound simplicity evolves while solving a multifaceted problem, such as the tepee, and sometimes it is forced upon us by severe constraints, as with the Conix One inhaler and Morse code. An elegant algorithm can be a simple formula that can solve a complicated problem through repeated use. These are just a few examples of inventions in which an ingeniously simple solution has achieved much from little.

The next section focuses on elegance in a related but slightly different way. Here I examine elegant inventions that are basic or "core" inventions. These are inventions that form the basis for many subsequent inventions. They are inventions of such a fundamental nature that their uses are both varied and numerous and go far beyond what the inventor could have imagined. While these inventions are also ingeniously simple, they can be used in many different applications. The wheel serves as a classic example. The characteristic of elegance highlighted in this next section is *doing the most*.

Core Inventions: From Screws to Microprocessors

Figure 6.7: The screw as a parking garage ramp. *Photo from iStockphoto.*

The devices that are classically described as simple machines—the lever, the inclined plane, the wedge, the pulley, the screw, and the wheel and axle—are all basic and simple designs that we take for granted today. But imagine for a minute that you were the inventor of the screw. Perhaps you knew about the inclined plane and had the breakthrough thought that if you curved the inclined plane so that it spiraled, you could get the same effect in a much smaller amount of space. As your thinking progressed around this invention, you realized that in its simplicity, there were many things that it could do. It could be used as a fastener or to transport materials; it could be made larger, so people could walk up a circular path to easily get to the top of a building; it could be used to pump liquids or to lift heavy objects. You realize that you invented something that has more uses than you can think of. It is a fundamental, simple, and elegant invention that can be modified or customized to fulfill many different needs.

This idea of expanding the uses for fundamental discoveries has had far-reaching consequences. For instance, is the creation of glass from sand an elegant invention? By my definition, it is. These days, we take glass for granted and don't really look at it as an invention, but the discovery of how to make glass was considered to be revolutionary. Today, glass is used in products from windows to fiber optics to computer chips, so the idea of transforming sand into a new state (probably discovered by accident) fits my definition of elegance. As with the screw, the vast array of uses and subsequent inventions derived from the creation of glass shows the extensive reach of this core discovery.

What about a brick? Before bricks were invented, structures were probably improvised using dirt, rocks, and trees. Bricks, made first of dried mud and later from kiln-hardened clay, literally form the basic building blocks for thousands of different structures. That the simple brick can be used to create structures of immense complexity and utility is a testimony to the elegance of this invention. One of the most popular and long-lived toys, the LEGO building block set, is based on plastic bricks held together by friction. Again, the elegance is how so much can be done with such simple elements.

Is a child's ball an elegant invention? I would answer yes. Hundreds of games can be played with a ball, some traditional and others made up on the spot. Yet a ball is merely a simple sphere. The idea of playing with a ball probably came from tossing around stones, pieces of wood, or even skulls of animals. So, although the modern ball is more a refinement than a new invention—with everything from a beach ball, baseball, basketball, and tennis ball among its many variations—it fits my definition of elegant because so much can be done with it, and it is in itself so basic.

A more modern invention in this category is the microprocessor. Building on the development of integrated circuits, the microprocessor is a miniature computer on a single silicon chip. Microprocessors are used in everything from washing machines to automobiles to personal computers. As we progress further into the twenty-first century, many more products will be designed with microprocessors to provide them with intelligence. As technology enables microprocessors to become smaller and more easily incorporated into different forms and materials, we will see new uses develop, such as smart clothing or smart building materials. The uses are endless, as microprocessors can be embedded into innumerable products to improve their functioning and will be used in ways far beyond the imagination of the original inventors.

Many inventions or discoveries that I am calling elegant are so ubiquitous that we don't think of them as having ever been created. Many were no doubt perfected over time with much iteration before they came into common use. These inventions are very basic, but their utility is so great that they have a "product life cycle" that has extended over thousands of years. They also continue to inspire derivative inventions.

Adaptive Inventions: The Elegance of Self-Regulation

This third aspect of elegance can be seen in inventions that mimic a living organism's ability to adapt to change. Inventions of this type embody a physical, biological, or chemical property that enables them to adapt to various conditions they encounter during use. These properties are intrinsic to the invention and serve to regulate the invention's response to its environment. Again, this is most easily seen through example.

A spring, as mentioned earlier, is a simple example of such a system. It regulates itself according to Hooke's law of elasticity. If you pull on a spring, it pulls back. Stretch it some more, and it pulls back harder. In fact, the force that it pulls with is proportional to how far you stretch it. In other words, it always wants to come back to its unstretched length. How is this a self-regulating system? The spring senses change from its original condition and reacts *proportionally* in a manner that will try to return it to its original length. The more its original condition is changed, the more strenuously it reacts. If a spring reacted with the same amount of tension no matter how far it was stretched, then it would not have this property of control. One of the reasons the spring is such a useful (and elegant) device is that it is self-regulating and automatically controls tension in response to its stretch or compression. Like the paper clip, the spring is a simple piece of wire, in this case, hardened and usually coiled. There are many variations of shape and design, but they all adhere to the basic property that the return force increases in proportion to elongation or compression. Within this simple system lies a sophisticated degree of control.

The solar chimney is another self-regulating invention that relies on the principle of convection. Solar chimneys are designed to cool buildings in lieu of, or as a supplement to, air conditioning. The solar chimney is a vertical shaft, extending higher than the roof of the building, that uses a black or glazed surface on its exterior to absorb the sun's heat. The bottom of the shaft is vented and located in an open area within the building. The warmed air at the top of the chimney rises, causing cooler air from below to be drawn up the chimney. This provides both ventilation and cooling to a building. The speed at which the hot air escapes is determined by the temperature difference between the air inside and the air at the top of the chimney. On hotter days, there will be a greater difference, and the system will work harder to maintain thermal equilibrium. The larger the temperature difference, the faster the airflow, and the greater the ventilation and cooling. Much like the spring, this system regulates itself based on its external situation; in this case, adjusting the rate of airflow in response to temperature.

Finally, the electric motor, like the spring and the solar chimney, is intrinsically self-regulating. If a motor is attached to a heavy load, it faces much greater resistance to turning than it would with a lighter load. The amount of torque (twisting force) a motor can apply in order to rotate its shaft is determined by the resistance of the load it is trying to turn. If a motor's rotation is slowed by a large load, it automatically increases the current that it draws, so that it can apply more twisting force in an attempt to increase its speed of rotation. As it overcomes inertia and its speed increases, it needs less twisting force, which it automatically reduces by drawing less current. In other words, it *automatically* regulates the torque it applies in response to the load it has to turn.

You may experience this effect when you turn on a large electrical appliance in your house—an air conditioner, for instance. As you switch on the air conditioner, you might notice that the lights in the house dim momentarily as the motor draws a large amount of current to enable it to start turning. Once the motor begins to turn, the current need is reduced. The amazing thing about the motor is that it is completely self-regulating. It alone monitors how much torque and, thereby current, it needs. The motor automatically makes adjustments as it regulates its needs to its external conditions.

How does this self-regulation work? Briefly, in accordance with Lenz's law of electromagnetism, whenever a motor spins, a current *in the opposite direction* of the incoming current is created through the motor's magnetic field. The faster the motor spins, the greater the opposing current. The net current drawn is equal to the incoming current minus the opposing current. If the motor is spinning fast, as it would under a small load, the opposing current is almost equal to the incoming current, and the net current is very small. However, if the motor is slowed by a large load or is just starting up, this opposing current is minimal. The net current is then very large, which makes the motor's twisting force very large. This self-regulation requires no extra parts; it is simply a function of the amount of opposing current generated as the motor's rotor turns at different speeds.

Without self-regulation, motors would need an array of sensors to manage current flow under different loads. The fact that motors automatically adjust torque to match the load shows elegance through the simplicity of an innate feedback system. In other words, by automatically reacting to its external load, the electric motor manages itself by adapting to changes in its environment. We can compare self-regulation in inventions to homeostasis in biological organisms. For example, our breathing adjusts to the amount of oxygen our bodies need at any given moment. If we are running, we breathe hard; in sleep, our breathing is slow. This ability to self-adjust is integral to both biological organisms and the inventions discussed above. This intrinsic ability to adapt to external changes in order to optimize function is a hallmark of elegant design.

Smart Inventions: Inventions That Change with Change

Invention can aspire to yet another level that goes beyond self-regulation. Instead of responding predictably as it interacts with its environment, this kind of invention can actually change in an unpredictable manner and grow as it is used. In order for this to happen, the invention must work in partnership with greater intelligence. In the examples that follow, that intelligence is provided by human beings. The invention harnesses this intelligence and acts to facilitate its expression.

Growth through Partnership

The Internet is the exemplar of an invention that changes with change. The outgrowth of more than thirty years of experimentation with computer-to-computer communication known as networking, the Internet allows anyone with access to a personal computer to participate in the development and exchange of information. What is amazing

Figure 6.8: The Internet connects the entire world. *Photo from iStockphoto.*

about the Internet is that it provides no information itself; it is merely a facilitator of contact and exchange. Its reach is worldwide; its growth unlimited. Anyone can be a node on the network by operating a server. Anyone can have a Web site and share whatever information he or she wishes with the entire world. The concept is astonishing.

The Internet is a communications skeleton that can support an unlimited amount of information exchange. This invention not only does a lot with a little, but the amount that it does is continuously growing. The hardware—consisting of servers that form network nodes and personal computers from which users gain access to the network—is modular: it can easily be connected to the network, just as a household appliance is connected to the electric power grid by being plugged into an outlet. The ease of setting up a server or personal computer is a crucial factor in promoting growth and universality. The Internet is one of those inventions, like the screw or the wheel, whose uses will continue to be discovered over the coming decades.

The human-machine partnership works so well with this invention because each role is clearly understood. The machine is a facilitator, and the human is the information provider and recipient. Interestingly, one of the key reasons for the success of the Internet is that it is not owned by anyone, nor is it designed to benefit anyone financially. The idea is to make it so inclusive that all who want to participate can. If its use were limited by a fee or exclusive membership, it never would have become the universal means of information exchange that it is today.

Wikipedia, the free Internet encyclopedia that anyone can edit, demonstrates another fascinating aspect of the Internet called open collaboration. Wikipedia describes itself as follows:

> Wikipedia is a multilingual, web-based, free content encyclopedia project. Wikipedia is written collaboratively by volunteers from all around the world. With rare exceptions, its articles can be edited by anyone with access to the Internet, simply by clicking the *edit this page* link. The name Wikipedia is a portmanteau of the words *wiki* (a type of collaborative website) and *encyclopedia*. Since its creation in 2001, Wikipedia has grown rapidly into one of the largest reference Web sites on the Internet.[4]

The idea of writing an encyclopedia using volunteers who are free to write and edit as they choose is a radical paradigm shift in method. The accepted model of how to accomplish large tasks that involve many people is through a centralized and hierarchical structure. The structure is triangular, with primary decision makers at the apex. The Wikipedia model is the opposite of this structure. While there are rules about the editing process and ways to resolve conflict, Wikipedia uses the Internet's wide reach and ease of communication to have a group of volunteers—people from all over the world who do not know one another—to work on a problem; in this case, an encyclopedia entry. The idea is that *anyone* can add or modify information (within the guidelines) and that the cumulative effort of interested volunteers worldwide will eventually result in an accurate and comprehensive entry. Most conflict is resolved by using a discussion page that is part of the overall entry. The Wikipedia entry becomes a living entity that can always be modified. With the exception of entries on controversial subjects that need to be more closely regulated by an editorial board (such as those dealing with political or religious topics), most entries go through a process of becoming more and more refined as time progresses. The editors of Wikipedia comment on the process of building an article and how it differs from the more traditional centralized process:

Figure 6.9: A Wikipedia puzzle symbol. The jigsaw puzzle design exemplifies the collaborative, dynamic nature of Wikipedia. *Image by Kimbar, http://commons.wikimedia.org/wiki/User:Kimbar/gallery.*

> As a wiki, articles are never complete. They are continually edited and improved over time, and in general this results in an upward trend of quality, and a growing consensus over a fair and balanced representation of information.

> Users should be aware that not all articles are of encyclopedic quality from the start. Indeed, many articles start their lives as partisan, and it is after a long process of discussion, debate and argument, that they gradually take on a neutral point of view reached through consensus. Others may for a while become caught up in a heavily unbalanced viewpoint which can take some time—months perhaps—to extricate themselves and regain a better balanced consensus. In part, this is because Wikipedia operates an internal resolution process when editors cannot agree on content and approach, and such issues take time to come to the attention of more experienced editors.
>
> The ideal Wikipedia article is balanced, neutral and encyclopedic, containing notable, verifiable knowledge. An increasing number of articles reach this standard over time, and many already have. However this is a process and can take months or years to be achieved, as each user adds their contribution in turn. Some articles contain statements and claims which have not yet been fully cited. Others will later have entire new sections added. Some information will be considered by later contributors to be insufficiently founded, and may be removed or expounded.[5]

This decentralized approach to building, made possible by the Internet and the Wiki software that runs Wikipedia, is elegant because, once launched, it fosters self-growth. The invention is focused on harnessing as much human intelligence as possible, without discrimination, in the hope that the human qualities of curiosity, creativity, insightfulness, and reflection along with accumulated knowledge will produce a superior end result. The success of this approach and the Wiki software is predicated on the wide reach of the Internet.

An invention that can continue to refine itself and thereby effect its own change mimics the elegance of a biological organism. Unlike the fixed response of self-regulating inventions, these inventions grow and modify themselves with use. Wikipedia is one such application made possible by the Internet. The Internet serves as a platform for many more such inventions that, in partnership with humans, will continue to create information structures that behave like living things.

Self-Reconfigurable Robots

Self-reconfigurable robots also take their cue from one of the fundamental principles of biological systems—the ability of basic modules (cells) to organize themselves into more complex systems. Self-reconfigurable robots are "able to deliberately change their own shape by rearranging the connectivity of their parts in order to adapt to new circumstances, perform new tasks, or recover from damage."[6] The elegance of such a system is that it consists of basic units or modules that can combine in numerous ways to perform different tasks. In other words, *doing a lot with a little.*

Figures 6.10: The M-TRAN self-configurable robot developed at Tokyo Institute of Technology and AIST, Japan. *Reprinted with permission of AIST.*

The modules have the intelligence to combine with other modules in a specific configuration that will allow the system to achieve an ultimate goal. Of course, the goal is predetermined by the human programmer, but the manner by which the modules assemble to reach the goal is not. For example, if the goal of a modular system is to recover from damage, it must first identify the damaged element and then figure out how to either replace it or function without it. Because the damaged element can be anywhere, the self-reconfigurable robot must have the intelligence to react to numerous possible situations.

Self-reconfigurable robots can assume different shapes to perform various functions. They can also use this "shape-shifting" ability, called morphogenesis, to move in different ways. The M-TRAN robot (shown in two possible configurations in figure 6.10) can move as a four-legged walker or unfold into a linear chain and move like a snake by undulating its body.

Aside from self-repair, morphogenesis, and adaptability to different environments and tasks, self-reconfigurable robots have been designed to replicate themselves. The Molecube robot, developed at Cornell University, is the first robot to demonstrate self-replication.

Figures 6.11: An individual Molecube and a Molecube robot reproducing itself (developed at Cornell University, by team members Zykov, Mytilinaios, Adams, and Lipson). *Photos courtesy of Cornell University.*

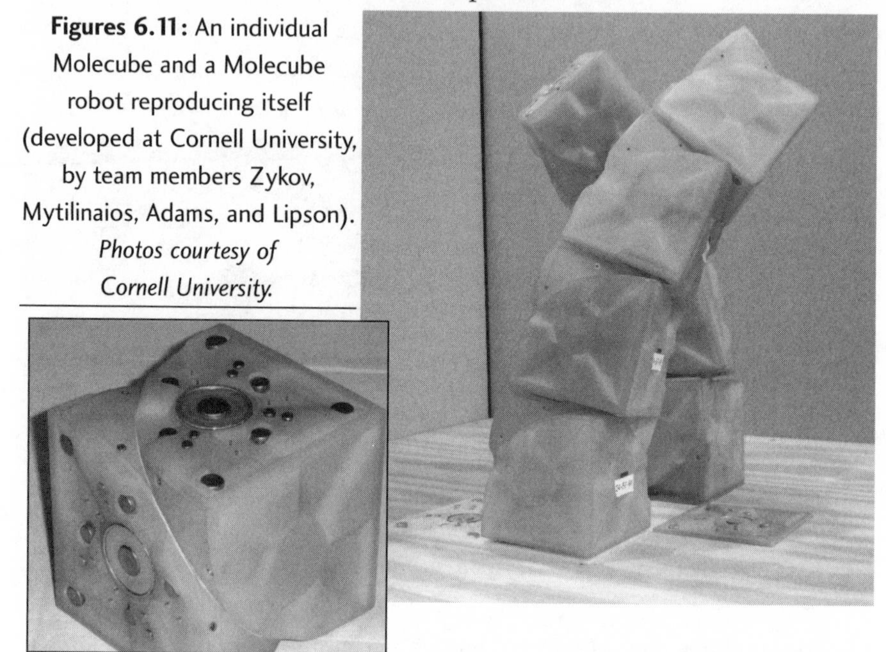

Starting with a stack of four cube-like modules, the robot reproduces itself using modules from its stack as well as additional modules. The first stack picks and donates modules to the second stack, which, as it begins to grow, works in conjunction with the first stack to complete its formation. The basic module is the single variegated cube, shown in figure 6.11, that contains a microprocessor and a motor. Each module can swivel on an axis and connect to other modules through an electromagnet. This robot provides an impressive demonstration of the possibilities of using self-reconfigurable robots. Is this the twenty-first century's intelligent brick? One can imagine robots that create their own structures or reconfigure themselves to adapt to new or changing environments. This adaptability gives these robots an elegance born of a simple module or cell growing to meet the needs imposed by its surroundings.

The ability of these modules to organize themselves to the task at hand shows the power of something that is generically simple over that which is very specific but more complex.

Malleability in Design

The final example of elegance is what I call malleability in design. This means that an invention is designed in such a way that it can be customized by whoever uses it without its overall purpose being sacrificed. Malleability is difficult to achieve. On one level, it can be argued that a basic control feature—the three-speed switch on a fan, for instance—represents a kind of malleability in the design of the product. But the malleability that makes a product elegant goes far beyond that small amount of customization. It is easy to see how malleability is an underlying theme of the Internet and software such as Wikipedia. Both of these inventions provide a structure (with a purpose) that is designed to be customized. Self-reconfigurable robots are also malleable, as they can assume many different configurations based on the task they need to accomplish. They are malleable because they consist of modules that can be assembled in different ways. Modularity is the most common way to make something malleable. The goal of a malleable invention is that each user can customize it specifically to his or her needs, while maintaining the invention's distinct purpose.

Software, for example, is very malleable at the level of programming. Modules exist in the form of libraries of functions that can be put together by the programmer. New modules or subroutines can easily be created. The extreme malleability of software in the hands of the programmer is the reason it is called *software*. Once a formalized program is created that will be sold to the nonprogramming public, the software becomes far less malleable. Wouldn't it be nice if the "soft" in software applied to the user of the product? There have been some attempts to achieve this, such as "plug-in" modules tailored to specific programs, but most customization is generally limited and superficial. Software is rarely sold as customizable components that can be assembled by users in a manner tailored to their exact needs. Instead, users can make some small modifications, such as turning various features on or off, but they cannot really customize the product to work in exactly the way they want.

However, there is one example of software that stands out for its malleability. That is the spreadsheet. Interestingly enough, the spreadsheet is not modular but rather provides a template for unlimited customization. It is designed with a purpose (manipulation or comparison of numbers) and can be customized to suit the specific needs of any user who wishes to examine numerical data.

Truly malleable design is difficult to find in manufactured commercial products. We can find it, however, in products like 80/20, the Industrial

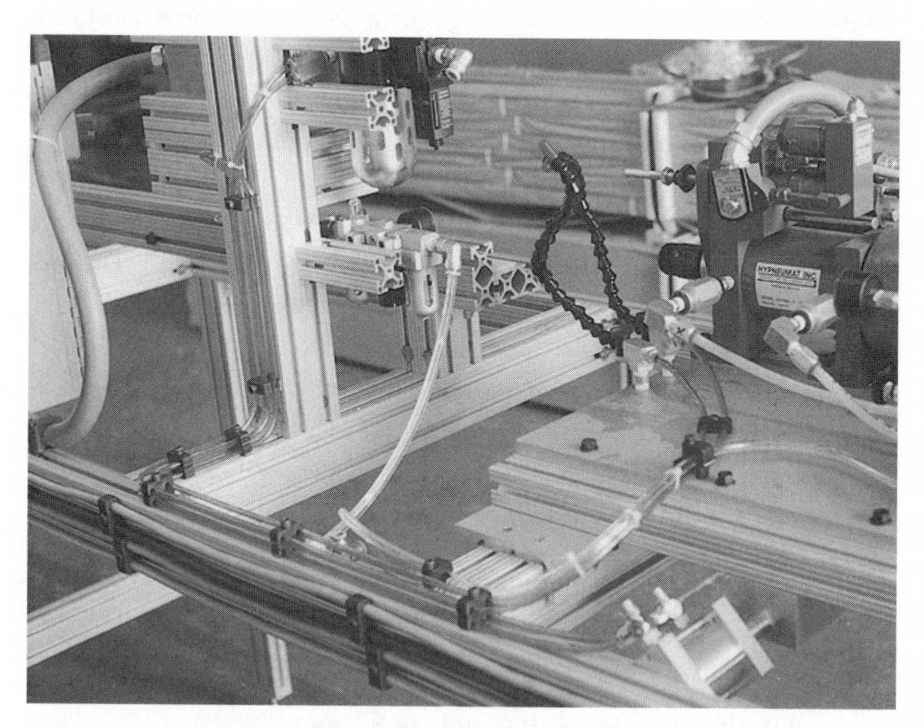

Figure 6.12: Machinery built using an 80/20 Industrial Erector Set framework.
Photo courtesy of 80/20 Inc.

Erector Set, which is a modular system of slotted aluminum framing used to assemble a customized framework for machinery or other applications. The 80/20 product fulfills the two major criteria of malleability: first, it has an overriding purpose, which is to provide industrial-strength structural framing, and second, it is almost infinitely customizable. A large number of unique designs for everything from warehouse racking to automated manufacturing machinery have been made using 80/20 modules as their basis. The flexibility to design customized structures that are strong and can withstand the rigors of an industrial environment make 80/20 an outstanding example of malleable design in the marketplace.

The Swiss Army knife and other tools inspired by its design attempt to produce multifunctional all-in-one solutions. An example of an eleven-function multipurpose tool, no larger than a credit card, is shown in figure 6.13. While this type of design can perform many different tasks, and in this sense is "customizable," it can be considered elegant only if (1) it can perform each selected function as well as a device solely dedicated to that function, and (2) the parts that are not being used can disappear without becoming a hindrance to the task currently being performed.

Figure 6.13: An eleven-function multipurpose tool small enough to fit in a wallet. *Photo by Ivan Boden.*

Ultimately, malleability means being able to customize the final product to your exact needs. Malleability is a major challenge for any inventor or designer, but when done elegantly, it gives the product tremendous flexibility—once again, *doing a lot with a little.*

Characteristics of Elegance—A Summary

We have seen that elegance can have many forms. In addition to the examples described, many of the inventions mentioned in previous chapters would also qualify as elegant. Characteristics of elegance include:

- Ingenious simplicity,
- Multiple and flexible uses,
- Functionality derived from intrinsic properties,
- Self-regulation, and
- Growth, customization, and development through use.

Many of the principles of elegance are drawn from natural paradigms. Biological systems have an elegance that has evolved out of necessity. Living organisms are designed to *do the most with the least*. More and more, as we strive for elegance in our inventions, we will use the natural world as a model.

Chapter 7

ROBUSTNESS

Along with simplicity and elegance, the third important element of good design is robustness. Robustness describes how reliably a device can function under adverse conditions. One definition describes robustness as "capable of successful misuse." A robust invention is designed to work despite problems. If failure is inevitable, the invention will fail in a controlled manner, such as a shatterproof windshield does. Robustness can be built into a product or system in many different ways, including:

- Strength
- Redundancy
- Simplicity
- Self-healing
- Managed failure

As we explore each of these methods, we will find that robust designs are often both simple and elegant. These three design ideals are closely related and often intersect. By pursuing one, you very frequently attain another. We will begin by examining what is perhaps the most apparent route to robustness: strength.

STRENGTH

Making something very strong is an obvious way to make it robust. Bridges, for example, are built to withstand several times their worst-case load. Airplanes are also designed to tolerate much more stress than they are expected to encounter. Most inhabitable structures are designed with excess strength and, in some cases, excess flexibility, so that they will not collapse under large momentary loads. How much strength is necessary for a structure to be robust? This is an age-old question for engineers. The answer often depends on how critical it is that the structure not fail. A bridge failing, for example, would result in loss of life. The same can be said for a building or a mechanical linkage in an artificial heart. Balanced against protection from failure is economic cost. Should all houses be built to withstand hurricane-force winds? Maybe on the Florida coast, but probably not in New York.

Robustness through strength includes the ability to withstand both extreme short-term loads and the cumulative effect of loading and wear over time that can cause a structure to fail through fatigue. Commercial airplanes, which experience many cycles of pressurization and depressurization, are designed with an understanding that materials will fatigue from the repeated stress. Part of the robustness of aircraft design comes from strict maintenance schedules that require that parts be replaced before they can fail.

The area of consumer products is one in which robustness through strength is often sacrificed to the competing value of lower cost. This was not always the case; prior to the 1960s, products were "built to last." However, three things changed this paradigm: First, price competition forced companies to look for ways to reduce costs; second, consumers were willing to accept products that had a shorter life in return for lower prices; and finally, industry realized that if a product had a shorter life, the consumer would have to go out and buy another one, thus extending the sales cycle. Both consumer electronics and automobiles are examples of products whose robustness has been compromised for the principle of "built-in obsolescence." If your computer doesn't break down in three or four years, you will still need to buy a new one because it will be obsolete. The upshot of this is that many consumer products are now designed to be just "good enough," rather than "built to last," which means that robustness through strength is pared to a minimum. As long as the product doesn't fail while under warranty, it is strong enough.

Adding mass or bulk is not the only way to make a product stronger. Walking the tightrope of strength versus cost can lead to some very innovative and inventive ways of achieving strength at low cost. New materials are a primary example. For example, carbon composites are replacing aluminum in airplane structures. These materials are not only less expensive, but they are stronger and more resistant to fatigue and wear. Design is another way to achieve robustness through strength. The simplicity and elegance of the truss and the arch are classic examples of design that does not rely on mass or excess material for strength. Through clever design, less can actually give more.

Making something strong is only one way of achieving robustness. There are situations in which strength isn't enough, and robustness needs to be ensured through redundancy.

REDUNDANCY

Redundancy involves having backup systems, so if one fails, another will take over. On the surface, it seems inefficient to do the same thing twice; however, if that thing is important enough, it might well be worth it. Backing up critical data on a computer, for example, is a task that no one would question. The prevalence of RAID systems in home computers, which contain two identical mirrored hard disks (one is an exact copy of the other) in case of disk failure, is testament to this need. While redundancy increases cost and the amount of hardware required, it also dramatically increases the robustness of any system. The possibility of failure in both the main system and its redundant backup is very slim. Like strength, the value of redundancy must be weighed against the cost of failure. In an airplane, redundant systems are mandated. If there is a failure in a critical part of the avionics, there must be a backup. The cost of failure can be human lives, and that certainly trumps the cost of a secondary backup system. In fact, the redundancy in critical cases such as aircraft typically requires a completely separate set of isolated systems that can be activated immediately in case of failure.

Redundancy is also what makes the Internet so robust. Nodes on the Internet are connected through multiple pathways, so that if one path is blocked, another can be found. The redundancy of this lattice-like distributed network is key to the functioning of the Internet. Paul Baran, an early pioneer of computer networking, developed this model based on the robustness of the human brain, which can create alternative neural pathways in the event of injury.[1] His concern at the time was building a computer network that could survive a nuclear war with the Soviet Union. He saw redundancy as the key. Baran concluded that a highly redundant system, in which each node connects to at least three or four other nodes, would provide enough alternative pathways between any two nodes so that the entire system would be able to communicate, even after a nuclear attack. Figure 7.1 illustrates the development of Baran's ideas from a centralized system to a decentralized system and, finally, to a completely distributed network.

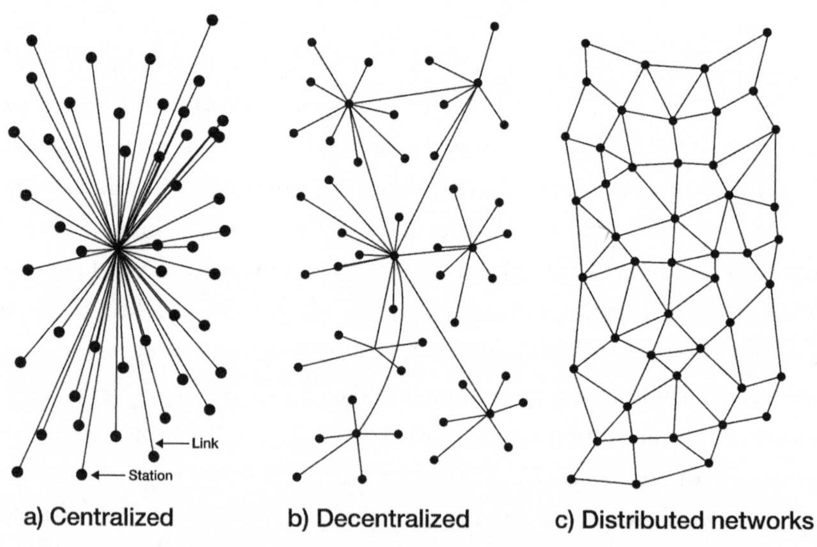

Figure 7.1: Baran's development of the distributed network model.
Drawing reproduced with permission from the IEEE (Paul Baran, On Distributed Communication Networks, IEEE Transaction on Communications Systems: © 1964 IEEE).

Today the Internet does not have the level of redundancy that Baran envisioned necessary to survive a nuclear attack. However, its architecture is based on the same concept of a distributed network with redundant pathways between all points.

Modularity is another concept that lends itself to redundancy. Self-reconfigurable robots, discussed in chapter 6, are robust because they are made up of many similar modules. If one is damaged, it can be bypassed or replaced by a duplicate. I will discuss this further in the section on self-healing systems.

Redundancy minimizes the chances of overall system failure by providing a built-in automatic backup. This creates a robust system that can tolerate failure of one or more of its parts. Redundancy becomes most important in very complex systems and in systems where failure has an extremely high cost. While redundancy can make a system robust by adding more, designing a system to have less is another way to reduce the chance for failure.

SIMPLICITY

Simplicity has been a theme throughout these chapters. We have looked at simplicity as a design goal and as a way of achieving elegance. It is also a way of making a product or system robust. The simpler a system is, the less that can go wrong; the more complex a system, the more that can go wrong. In mechanical designs, simplicity can be achieved by reducing the number of moving parts. In a structural design, simplicity can be achieved by using a minimal number of structural elements to provide strength. Sometimes a simple solution might be to allow a structure to flex or move instead of shoring it up. Simple designs use their surrounding environment in a productive way. The ridge and soffit vents or the solar chimney, both of which use convection, are examples of very simple and robust methods of cooling a house. They use their natural environment as their energy source, have no moving parts that can wear, and are completely self-regulating. As long as their air paths remain clear, they will continue to function.

Sometimes we can use the natural properties of a material in a way that gives us a simple and robust solution. Metals, for instance, expand when heated. This enables us to use temperature to cause motion. If strips of two different metals are attached to each other, the difference in expansion when heated (or contraction when cooled) will cause the attached metals to bend in one way or another. If the metal strip assembly is placed between electrical contacts, it can serve as a temperature-sensitive switch. The bimetallic switch consists of two pieces of dissimilar metal bonded together. This switch is found in thermostats, hair dryers, coffeemakers, and many other devices that need to be turned on or off in response to their temperature. The robustness of this solution is that there are no moving parts; the metals bend within their elastic limits and therefore can go through millions of cycles without failing.

A more homey example is the whistling teapot. A simple slit or hole in the spout of the teapot causes a whistling sound when the water comes to a boil, signifying that the water is ready for use. Simple, ingenious, and robust. Here, the escaping steam—a natural consequence of boiling water—is simply harnessed to perform a signaling function. We should all invent something so clever and robust.

Simple, robust solutions typically harness a force of nature, such as gravity, or a property inherent in a material.

SELF-HEALING

Self-healing systems achieve robustness through their ability to repair themselves if damaged. These systems are modeled on self-healing biological systems. The robustness of the human body is due to the fact that it is constantly undergoing self-repair. Even parts that are not broken are regularly replaced. We are continuously shedding skin cells, for instance, and new ones are being manufactured. When injured, the body immediately responds with an effort to repair itself.

This model of self-repair is being adapted into machines and materials in order to increase robustness. Certain polymer materials can be engineered to repair themselves. These polymers, which have the ability to change state and restore their cross-links after being damaged, are being used to make self-healing materials. One example of how the effectiveness of these materials will be tested is an experimental self-healing house being constructed in the mountains of Greece. This area is prone to earthquakes and tremors. With the assistance of the NanoManufacturing Institute of the University of Leeds, polymer nanoparticles will be embedded in the walls of the house. When the walls come under stress from forces that could cause them to fracture, the nanoparticles will flow like a fluid into any small cracks that form. The fluid will flow until the stress abates, and then, after filling the cracks, the polymer will revert to a cross-linked solid. This allows the walls to repair small cracks within themselves in an effort to prevent a more catastrophic failure.[2]

Another example of a self-healing system under development is that of carbon-fiber composite materials. These materials are used in many modern structures from racing boats to bicycles to jet aircraft. Carbon fiber has the advantage of a very high strength-to-weight ratio. However, one problem with these materials is their tendency to separate internally or delaminate when large impacts cause cracks to form parallel to the surface of the material. Researchers at Switzerland's École polytechnique fédérale de Lausanne have developed a novel technique to make this extremely useful composite material self-repairing.[3] The technique involves impregnating the composite with hundreds of small bubbles filled with liquid monomer molecules and smaller bubbles filled with a catalyst. In addition, actuators made of shape-memory alloy—a type of wire that contracts when an electric current is passed through it—and sensors made up of optical fibers are embedded in the composite mesh. When the composite is subjected to a strong impact, the optical fibers in the area struck are compressed, dimming the light that passes through them. In this way the specific area of damage is sensed. The shape-memory wires are then triggered to contract as the monomer and catalyst bubbles are ruptured due to the shock of the impact. This effectively reduces the size of the delaminating crack and allows a small amount of monomer to flow into it and repair it. This process is analogous to gluing and clamping a cracked piece of wood.

This process might sound complex and futuristic, but the development of "smart" self-healing materials with damage-sensing logic devices is currently being pursued in laboratories throughout the world. The tremendous financial and practical benefits of robust self-healing materials will continue to drive this research forward. Eventually, self-healing materials will be commonplace in all failure-prone structures.

In chapter 6, I described modular robots that have the ability to replace defective modules. A novel four-legged robot developed at Cornell University is designed to be able to suffer an extreme injury, such as the loss of a limb, and figure out how to compensate so it can keep moving. The robot, called Starfish, uses an algorithm that determines the optimal use of limbs regardless of configuration. In other words, if one leg is broken off, the robot's controller computes how it can best move forward using only the other three legs. While this system cannot repair damage to itself, its software is designed to compensate for the injury so that the Starfish can continue to operate. The onboard computer figures out which of about one hundred thousand possible arrangements of the machine's parts will produce optimal locomotion. It then reconfigures the Starfish's gait so it can continue to move. The Starfish is an example of self-healing by compensating for, instead of repairing, an injury.[4]

Figures 7.2: The Starfish robot developed at Cornell University by Joshua Bongard, Victor Zykov, and Hod Lipson. *Photos courtesy of Cornell University.*

The Starfish robot uses software to model a physical entity and control its locomotion. Software systems are another area in which self-repair is increasingly being built in. As we become more dependent on software-based systems, their ability to diagnose and repair themselves on the fly becomes critical. Commercial software has a well-deserved reputation for failing in unexpected ways from which it cannot recover. As this industry matures, we should see more robustness in the form of self-healing and self-correction designed into these products. Throughout the evolution of Microsoft operating systems, from Windows 95 through Windows 7, there has been a somewhat successful attempt to design in robustness. Microsoft faces a similar challenge to that of the makers of the Roomba robotic vacuum: how to make their product work in unknown circumstances. For the Roomba, the unknown circumstances are room layout; for the Windows operating system, the circumstances are the wide variety of unknown software that will be loaded onto any given system. The interaction between various pieces of third-party software is almost impossible to anticipate. Microsoft added a feature called "System Restore" to Windows XP and subsequent operating systems, which allows users to reset the operating system to its prior settings and conditions. This creates a partly self-healing system. If the system is damaged, it can be healed by resetting it to a time before the damage occurred. This does not preclude the loss of data, but it does secure the basic system.

I believe that self-diagnosing and self-healing systems will become more prevalent in the future. Our dependence on advanced technologies requires that they achieve a level of robustness that will allow them to function regardless of challenge or internal failure.

Managed Failure

A final way of achieving robustness is to assume that failure is inevitable and to design with that in mind. How can something be robust if it is destined to fail? That depends on the goal of robustness. We will look at two kinds of failure—the kind of failure from which a system can recover and the catastrophic kind of failure from which there is no recovery.

The goal of a system in catastrophic failure is often protection. A fuse is an example of a device that through its failure protects a system from damage. The fuse is damaged permanently, but its addition to a sensitive electrical system makes the system more robust by ensuring that critical and more expensive components will remain intact. The automobile provides many examples of sacrificial systems that are designed for catastrophic failure with the goal of protecting the occupants. Shatterproof windshields, mentioned at the beginning of the chapter, are designed to break under the force of impact without shattering shards of glass. Crumple zones are designed to fail by absorbing the energy of a collision, permanently deforming the automobile in the process, with the goal of protecting the passengers. In all these cases, robustness is achieved by sacrificing the less important for the protection of the more important.

The second case of managed failure is one in which failure happens as a part of regular operation. Information flow through computer networks is an example of this. Information is sent in coded packets, and occasionally those packets don't make it to their destination. They can collide with other packets or become distorted. The underlying assumption that makes network communication successful is that the communication will fail. The system is designed to assume failure unless the sender receives confirmation from the receiving computer that the information packet arrived intact. If no confirmation message is returned, or if the confirmation message fails to reach the sender within a certain time, the packet is automatically resent. Such a system assumes that failure will happen on a regular basis and includes recovery as a fundamental part of its design.

Yet another type of managed failure is referred to as *graceful degradation*. This is a case in which failure is managed so that, rather than being catastrophic, the effects are compensated for, and the system can continue to perform, albeit in a compromised manner. The Starfish robot's response when injured, as described in the previous section, is an example of graceful degradation. Another example is the failure of a software application such as a word processing program on a personal computer. Occasionally, the software fails in such a way that the program is forced to shut down. Normally, all unsaved data would then be lost. A program that degrades gracefully would save the data that would otherwise have been lost during a sudden shutdown in a recovery file. This file can be opened the next time the program is run.

In his book *Robot Programming: A Practical Guide to Behavior-Based Robotics*, Joseph Jones discusses the application of graceful degradation with regard to the design of sensors on robots.[5] He describes the failure of robot sensors to detect obstacles, which could result in the robot colliding with an obstacle and becoming stuck or damaged. His example of graceful degradation assumes the failure of the primary and even secondary sensors for detecting obstacles. In designing a robot for graceful degradation, there are multiple sensors that can cover the same ground. The primary sensor is the most powerful but also most prone to failure. We move down the hierarchy of sensors to the least powerful yet most reliable. Even though the primary sensor might fail, compromising the performance of the robot, the robot will not fail catastrophically but will degrade gracefully and continue to function. Managing failure makes a system more robust by assuming that failure will happen and by designing in the means to manage it once it happens.

In examining the various aspects of robustness, we can see that the design of a robust system requires an awareness of failure right from the beginning. Robust systems need to be made that way as part of the initial design, rather than "bulked up" later in the process. As I mentioned in the beginning of the section, robustness often comes from the *simple* and *elegant*. While these three principles are ostensibly independent, the combination of any two of them often results in the third.

PART III

MAKING IT HAPPEN

Chapter 8

PROBLEM SOLVING AND ITERATION

"Many of life's failures are people who did not realize how close they were to success when they gave up."
—Thomas Edison

"The devil is in the details."
—Anonymous

It can be said that there are two parts to inventing: the abstract and the nitty-gritty. The abstract is the creation of a novel idea as it exists in the mind of the inventor. The nitty-gritty is how to turn that idea into a working reality. Having an idea and giving the idea form—whether a simple sketch or an actual prototype—is the first step. Once you have something to work with, how do you proceed?

One of the great things about the brain, especially in a creative state, is that it smoothes over all the bumps and potholes along the way to an idea, preventing complicating details from getting in the way of the larger vision. If you focused only on the details, you would never be able to conceptualize the big picture. However, once you envision the big picture, the details must be examined. In this chapter, we will look at some techniques for dealing with the details—the problems, the frustrations, and the seeming impossibilities that often cause the inventor to quit or give up on an idea when he is just short of success.

A Systems Approach: Connect the Big Pieces First

A systems approach is a "top-down" method of problem solving that looks at the major systems or functional elements of the invention before focusing on the smaller details. The process is similar to writing software, where one charts out the major functional areas in very general terms and then logically connects them—all before writing a single line of programming code. Using a flowchart or block diagram can be helpful to the inventing process. Visual aids enable us to look at the major elements of our invention and ask how they will operate in relation to one another. Can all the systems work together as a whole?

For example, suppose your invention is an electric robotic lawn mower. Let's assume that the mowing mechanism is an off-the-shelf subsystem, and your invention is to turn the mower into an "intelligent" robotic device. Your first job is to identify the major functional elements. You will need:

- A mowing device
- A means of locomotion
- Sensors
- A processor
- A software program to perform the tasks

Each of these elements needs to work together. By diagramming the system in a top-down manner, you can see how they need to interact. In its early stage, your diagram might look like figure 8.1.

From your drawing you can see that the mowing mechanism will be independent of the other systems. In other words, once you turn it on, it will remain on and will not interact further with the other systems. In your initial conception, the purpose of the other systems is primarily for navigation and obstacle avoidance. As the design progresses through iterations, you might want to connect the mowing mechanism to the other systems. For

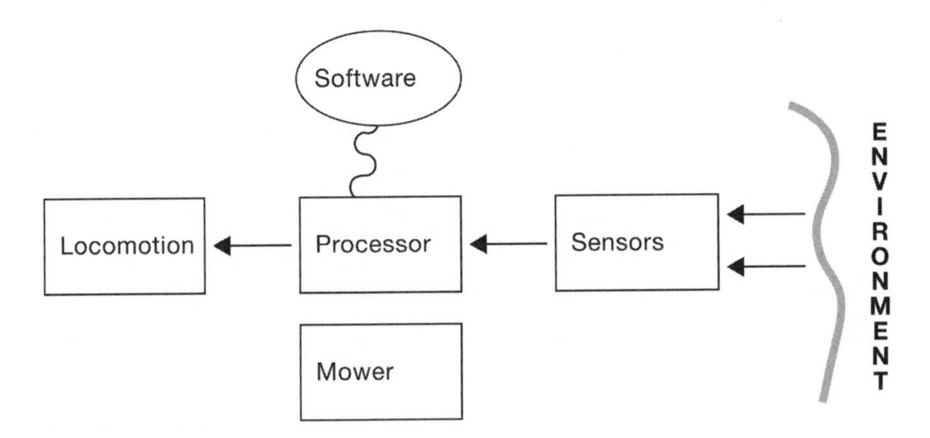

Figure 8.1: A block diagram showing the relationship between major systems for a robotic lawn mower.

example, you might want the processor to respond to a low battery charge or mechanical jam by turning off the mower. Regardless, the idea is to start simply and progress to a more complex design. You can see from your drawing that there are relationships among the sensing, processing, and locomotion systems and that all these systems will be controlled by your computer program. In fact, the software will be the key to this invention.

Now that you have defined the systems and their interactions in the most basic terms, you can go to work. Remember—and this is of key importance—*perfection is the enemy of the good.* In other words, your first pass at filling in the details will not reflect your ultimate design. Your primary goal with the initial design is proof of concept, getting the most basic form of your invention to work. You are creating a solid and stable place from which you can branch out and explore.

KISS: Keep It Simple, Stupid

Some of the best advice on how to begin any project is contained in the adage KISS (Keep It Simple, Stupid). Once you have diagrammed your systems, the next step is to get them to work on the most elementary level.

At this point, complexity is your enemy. You want your invention to be built and to function in the simplest manner. Remember, your first pass is a proof of concept. Can it possibly work? By definition, an invention is something that hasn't been made before, so who knows if it can work? That's why you want to keep it as basic and simple as you can.

Previously, we discussed simplicity as a design goal. Here, simplicity is both a design goal and a survival goal. For your invention to survive its proof of concept, it has to be made as simple as possible. Later, you will be able to refine it. A good analogy to this process is that of a rock climber climbing up a sheer rock cliff. Clearly the climber needs to progress in stages; he cannot bypass the rock and simply jump to the top. He first climbs a certain distance, then, after using a piton or bolt to secure his position, he begins the next segment of the ascent. It is an iterative process, with each step building on the last. This is the approach you will use to realize your invention.

To make your robotic mower as simple as possible, it needs only to be able to mow and maneuver in response to obstacles. The final locomotion system might be completely different from what you design in the first pass. The software will certainly grow from your initial test program. The sensor array will doubtlessly be expanded and modified to handle the myriad of situations that you could not anticipate at the beginning. The truism of the thousand-mile journey beginning with the first step is just this: The first step is the hardest, which is why we strive for bare-bones simplicity. Just make it work. From there, you have set the direction and can move forward.

Testing, Testing, 1, 2, 3

Once you have built a basic representation of your idea, the hardest part is over. You have overcome the barrier of moving from an idea into something physical, something that functions. But what you have built is still far from your final invention. It might be a series of components or mocked-up models connected with a rat's nest of wiring, but you have taken the first major step. It's there in front of you. Now leave it for a day or so and do something else. You need to come back to it fresh.

Once you return to that ugly, newly hatched duckling of your brilliant idea, don't be disappointed. What you see in front of you will bear little resemblance to your original idea, and you might be tempted to walk away from the project. Resist the temptation. You have completed your first and most difficult step. Now it is time to move forward. Begin by testing your creation. If possible, test one piece at a time. You are looking for basic functionality. Do the motors run when they should? Do the sensors sense? Can you make each system work on its own?

The next step is making the systems work together. Do the systems communicate effectively? Can you perform some very basic tasks that require the systems to work together? This will probably take some time; communication between systems often does not work right out of the box. Take your time debugging until you have achieved a satisfactory level of robustness. As robustness increases, make the tasks more complicated. Testing the systems together lets you see not only how to make them work but how they will *never* work, and what changes you will need to make as you progress.

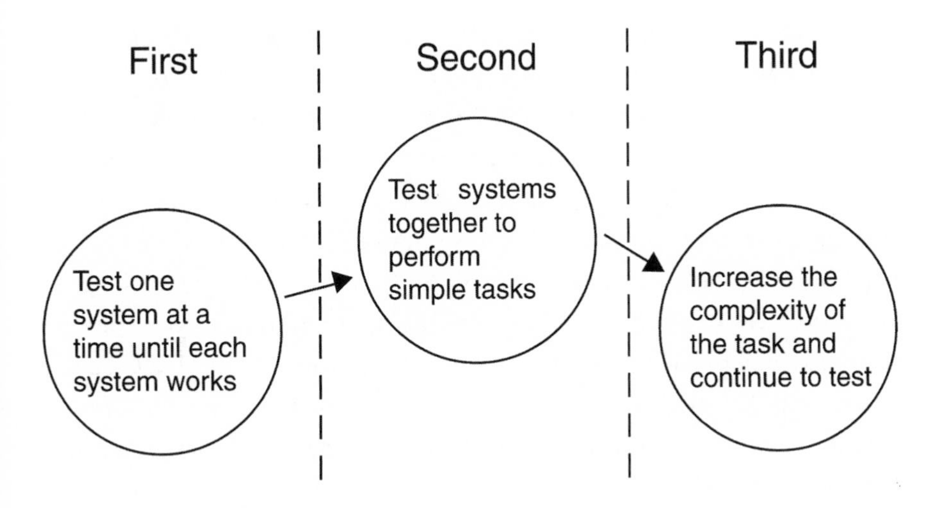

Figure 8.2: The three stages of initial testing.

Not all inventions are composed of multiple physical systems like the robotic mower. What if the invention is a financial product? Or a new food found in nature? Or a new method of making a useful chemical substance? While each of these might be a single, indivisible invention, they can only be successful if there are many working systems in place behind the scenes. For example, the key underlying systems for a financial product are its economic model and the business model for investment. For a new food found in nature, the systems include harvesting, packaging, transportation, and distribution. A chemical product might require procurement of raw materials and the ability to scale up manufacturing to a bulk level. No matter what the invention, there are systems that need to be developed. The process is always the same: Start at a simple and very general level and move on from there.

I will continue to focus on engineering inventions in this chapter, but these basic ideas can be applied to almost any kind of inventive activity.

What You Couldn't See from There, You Can See from Here

Now that the invention is in front of you, you can look at it from a different vantage point. What you couldn't quite see before is now much clearer. The challenges that lay ahead, the mistakes in concept, what is needed or not needed—all these things are now more obvious. You are still looking at the big picture, but now you are looking from a vantage point from which you can test your initial assumptions. Now is the time to evaluate the invention as a whole. Can you imagine it performing as desired? Are there areas of weakness or areas of potential failure? Do you want to make any major changes to the concept? Don't worry if you don't see the answers yet. You will revisit this process many times. The process of realizing and perfecting an invention is an iterative one. You proceed step-by-step, and with each step you gain new insight.

Let's return to our example of the robotic lawn mower. Assume that the mower is now constructed and is complete with locomotion, an array of sensors, and a controller. You have programmed the controller to perform rudimentary tasks that test the sensors and the motors. Now you have something that you can examine to see whether it meets your requirements. For example, looking at what you have, you can ask questions about the mower's ability to function: Is the sensor layout sufficient to detect all obstacles in all situations? Is the shape of the body optimal for navigating all types of terrain? Are there parts that will get caught on something and cause the mower to become stuck? Do the motors have sufficient torque? These and many more questions can be posed, addressed, and tested at this point. Remember, you have yet to design the details; you have not yet written a detailed computer program with navigation algorithms, nor have you done anything else that will make this invention completely functional. You are still at the concept testing stage, asking the general questions.

At this point, you might make large design changes. You might, for example, completely redesign the body of the mower because you realize the shape is not suited to the task. It's good to know these things before digging into the details of designing the subsystems. In a way, the purpose of iterative design is education. As you build, you learn; and then you go back, rebuild, and learn again.

Attacking from the Bottom

Once you feel you have exhausted the view from the top—that is, the creation of basic functional systems that give embodiment to your idea—it is time to flip things around and build from the bottom up. Now you can begin the detailed design of a small part or subsystem. This requires a very different approach than you used in the top-down process. Instead of focusing on the whole, you are taking a small piece and developing it in detail. This might be a piece of the software or some small functional element that is critical to the design. For the robotic mower, it might be the navigation algorithm. By changing your perspective from the macro to the micro, you also accomplish something that is not so obvious: you let your unconscious mind process what you have already completed. When you are ready to go back to the macro scale, new ideas will inevitably pop into your head.

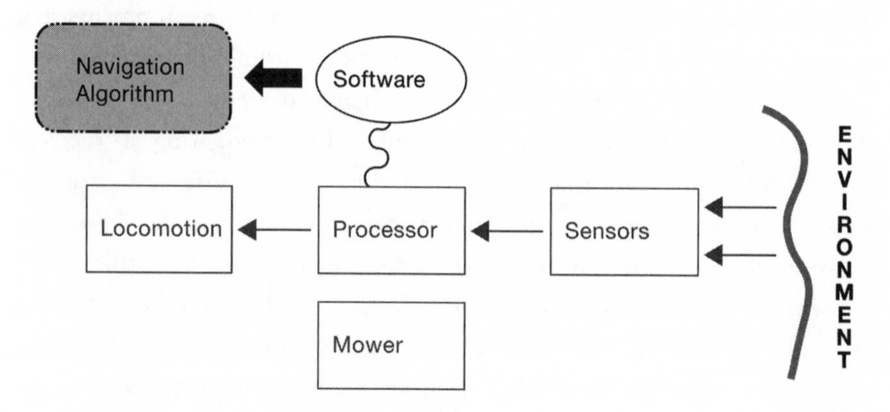

Figure 8.3: Switch focus to detailed design; in this case, the navigational algorithm.

Looking in detail at a small piece gives you the opportunity to focus on perfecting a part of the invention in isolation. The small pieces are the critical links that must work optimally for the invention to function. For your mower, it doesn't matter that all the hardware works perfectly if the mower can't reach every corner of a lawn. For a new food product, it doesn't matter if the distribution network is efficient and far-reaching if the food can't be harvested effectively. Each of the functional areas that are illustrated as boxes in a top-down design must gain lives of their own. Eventually, they will determine success or failure. Greater focus on the particular brings greater clarity to the overall picture.

Pick one area to start, and then dive into the details. Work in this area until you are confident that you have gone as far as possible. Once you have worked out a piece or subsystem in detail, take a break. Then switch gears again and go back to the macro scale, looking at the invention as a whole. How does your fully developed subsystem fit in? What changes are needed to accommodate it? Did you learn anything in designing the subsystem that will cause you to rethink the overall invention?

The change of focus from large to small and back again stimulates your brain to unconsciously process and synthesize your ideas. In chapter 3 we learned about feeding the brain and then letting it digest and work. This is exactly what you are doing here. You need to trust that your unconscious will be working on one problem even as you are consciously working on another.

Once you are confident that the developed subsystem can work with the invention, move on to detailed development of the next subsystem. Repeat the process of diving into the details and then returning to the large picture. This time, the challenge is a little greater, because your next subsystem has to work with the one you have already developed. While the bottom-up process of focusing on one piece and developing it in detail is crucial for each piece, you always have to keep in mind that all the pieces need to work together.

INVENTING WHILE YOU SLEEP

As was discussed previously, new ideas and answers to problems don't usually come to us on demand. They are the products of the unconscious—which works at its own pace—and they typically come into your conscious thoughts at times when you are not thinking about the problem. This can happen as you are waking up from sleep or when you are absorbed in other thoughts. You might be driving or performing some other task, not thinking very hard about anything, when an idea unexpectedly pops into your head. Suddenly, the solution to a vexing problem with your invention becomes obvious. "How did I think of that?" or "Why didn't I see that before?" you wonder, especially when you felt frustrated and thought you had come to a complete dead end.

This is how the process works. Your unconscious mind is at work on your problem even when you are not consciously thinking about it. The process of taking your invention from an idea to a working prototype is similar to that of imagining it in the first place. Both processes require you to feed your brain information from many different perspectives, both detailed and large scale. As you consciously put together the pieces, your unconscious will be doing the same.

Arbitrate and Iterate

As you move between detailed design and overall conception, conflict will invariably arise. As the master of the grand concept, you will say, "I want it this way. It must do this." On the other hand, as the design engineer who must deal with the details, you will say, "This is what I can reasonably achieve." Arbitrating between these two impulses is an opportunity for creativity, as opposed to mere compromise. Perhaps there is a third, better choice that can satisfy both sides. As you move between detail and grand concept, the design will inevitably change from the way you originally envisioned it. Each change brings a new perspective. Both concept and design grow through iteration. At first, the changes are large and sweeping, but through successive iterations, the changes become smaller and more refined.

Let us return to your robotic mower. Once you have created the navigation algorithm, perfected the sensing system for obstacle detection, and worked through all the electrical and mechanical requirements, you can proceed to testing by seeing how the mower performs in various situations. For instance, you can test it with different obstacle configurations or on a variety of lawns with differing terrain and see where it runs into trouble. This gives you the information you need for further iteration.

Limiting the Scope

Sometimes a key to solving a problem is limiting or redefining the scope of the problem. This might seem like cheating, but it really isn't. In chapter 1 we discussed the importance of defining the problem correctly. Sometimes the problem becomes clear only once you are working on the solution. For example, solving world hunger through sustainable farming might give way to teaching how to sustainably farm a specific crop in one particular geographic area. Another example might be a company that wants to sell a product globally; it might need to simplify its logistics by limiting its initial launch to a local area before going national or international. Or you might decide to limit the scope of your mower project to relatively smooth terrain, thereby eliminating the complex engineering and cost required to navigate rocky or muddy areas. Once success is achieved within a limited scope, it is easier to enlarge that scope. The problems that cannot be anticipated can be handled more easily on a smaller scale. Limiting the scope does three things:

1. It concentrates effort on a smaller target.
2. It increases the probability of success.
3. It provides an achievable goal for success that can be expanded at a later stage.

This is a case of trading quantity for quality. You can limit the scope at any time. It is not something done arbitrarily or due to laziness; it is a decision that is made after much effort and is based on newfound clarity—the clarity that success can be more definitively achieved by limiting or redefining the breadth of your solution.

STOPPING

How do you know when to stop? When is the invention done? The simple answer is never. In mathematical terms, the inventive process is asymptotic. This means that while the result is never completely perfect, it reaches a point of diminishing returns. The large picture and the small details eventually approach each other. As you alternate between thinking about the overall conception of your invention and perfecting the details, you will reach a point where the large and the small become nearly one and the same. Your invention and the vision behind it have come together. Almost. It is time to stop, or at least step back, once you feel that you have solved your problem in a manner that is simple, elegant, and robust. Sometimes, immersing yourself in one project suggests a completely new problem that needs to be solved. That's fine, but this new problem is also a new invention. Remember that you set out with a particular problem, and the job of your invention is to solve that problem. Again, remember that *perfection is the enemy of the good.* This means that the proper place to stop is short of perfection, because perfection is not achievable. Think of the asymptote. Once you are there, it is time to stop.

One more thought on stopping. It is psychologically difficult to end something, especially something in which you are very much invested. One way to get around this is to look to the next steps in the inventive process. Yes, you are stopping here with your invention, but now it is time to see how it will perform in the real world. You will have the opportunity to go back to it and revise it based on its performance as it is tested. It is not gone from your life. The process is not yet over.

PUTTING IT ALL TOGETHER

Once you have developed your invention into a working model that addresses all the issues that you can imagine, you are ready for a reality test. The reality test involves testing your invention in a controlled but realistic manner. This necessitates having people unfamiliar with the invention use it, evaluate it, and provide you with feedback. These people can be friends or colleagues or even people recruited through advertisement. You should give them basic instructions at the beginning of the trial, but make sure that you are not coaching them along the way. The more people, the more data, and hence, the clearer the analysis afforded by that data. Your own closeness to the invention can easily lead you to overlook significant flaws that this kind of testing will reveal.

With these first small-scale tests, you are letting your "baby" out to crawl in a new environment, and you will watch carefully. Be sure not to be overprotective or overly defensive. Your job here is not to defend your ideas or invention but to learn about potential flaws. Writers have editors for this reason. When you are too close to a subject (which you need to be to create and develop it), there are inevitable areas of blindness. Your job is to be a careful observer and listener. Gather all the data you can in an impartial manner. Interview your testers to elicit detailed feedback. Absorb it. Digest it. Then go back to reinvent your invention from a new perspective.

Anticipating the Unanticipated

The devil is in the details, especially those details that are difficult to anticipate. Large companies put their products in so-called beta testing for months to identify unanticipated flaws. Typically, the invention or product is released, on a trial basis, to a large group of people to use as they would the final product. They usually receive the product at no cost, with the obligation to use it in all possible ways and report back on its performance. The purpose of this kind of testing is to uncover flaws in the invention that a small group, using the invention in a directed and logical manner, would not discover. Most inventions are not put through this kind of testing until they are past the invention stage and being prepared for commercialization. However, it is worth mentioning that this is a tried-and-true technique for discovering flaws that are difficult to anticipate. If we look at large-scale projects, such as the development of a new operating system for the personal computer, we see that even after a long period of beta testing, flaws continue to be discovered upon product release. The more complex the product, the greater the difficulty of discovering all possible flaws. Flaws that occur over time due to wear or degradation are also difficult to uncover in testing. These flaws can sometimes be predicted using mathematical models, which can simulate effects over time. Again, this is more appropriate for the product development phase, which follows invention.

Documentation

Documenting both the inventive process and the final invention is very important and is often overlooked. We have previously discussed the use of a design log as an aid to creativity in invention. This is not only important as a way to explore ideas, but it also has legal value when applying for a patent. Perhaps not as well appreciated is the importance of documenting an invention in the later and final stages. Our memories are only so good, and with the passage of time, details escape us. Therefore it is important to create a reference document for an invention. This document will include both a complete description of the final invention as well as key worksheets of ideas, trials, data, and prototypes leading up to the final design. If the invention is an engineered device, you would include engineering schematics, software printouts and flowcharts, data sheets on key purchased components, a written description of the final invention as well as drawings, sketches, photographs, references, a detailed parts list including place of purchase and cost, operating instructions, cautions, environmental testing and failure analysis results, and anything else you can think of that will jog your memory about important details. Imagine that you need to reconstruct some key elements of the invention five years from now. Your documentation should be sufficiently detailed to enable you to do so.

The documentation also gives more value and substance to the invention. If you want to sell the invention or pass it on to someone else, the documentation provides necessary background. You might also want to include a detailed problem statement to put the invention in context.

Finally, you never know how your invention might fit in with some future need or application. The better the documentation, the easier it will be to show an interested party how the invention works in conjunction with that future need.

SUMMARY

First and most important, success doesn't come in one giant leap. As I have emphasized, it is an *iterative process*. We take a step, look around, assess, and then go on to take the next step. No matter how many steps you have taken, you can always go back. Often you need to progress to a certain point before basic truths become clear. The process of attacking a problem, backing off, changing perspective, attacking another aspect, testing and iterating, makes the problem manageable and clarifies its most critical aspects. The process of problem solving involves gradually digesting the problem until it is completely internalized. A great invention is not merely an intellectual exercise; it is a personal journey that requires the necessary effort and commitment to make it a part of you.

This journey from idea to concrete reality is often fraught with great frustration. I began this chapter with Thomas Edison's quotation "Many of life's failures are people who did not realize how close they were to success when they gave up." There will be many points during the process of turning your idea into something real at which you will want to give up. There will be wrong turns, times when the task ahead looks impossible, times of questioning your original assumptions. All this is included in the price of admission. Ideas are easy. Invention is work. Sometimes it is a good idea to step away from your project. Let it go for a short time. Walk away and forget about it. Your unconscious will continue to ruminate on the problem, and when you next approach the project, be it in a day or a week, you will find that what was previously insurmountable is now easily managed. This is part of the process. The important thing is to realize that there will be setbacks, and that you must not be overly discouraged by them. The ability to do this comes with experience. The first invention is a roller coaster of despondency and exhilaration. If you survive and are willing to do it again, future inventions are less emotionally volatile as you start to understand and anticipate the process.

Chapter 9

THE BUSINESS OF INVENTION

This is a book about creativity and invention, not about how to make an invention profitable. Nonetheless, the question of how to make money from an invention always seems to come up in any discussion of inventing. As soon as you mention that you are working on a new invention, you are asked about its potential in the marketplace. Entire books have been written about how to create wealth from an idea or invention. I will merely be touching upon the subject in this one brief chapter. I believe that the inventor invents not for financial gain (although that is a nice by-product) but because he is driven to do so, much like the artist who is driven to create art or the composer who is driven to create music. This said, there are ways to profit from invention. As in almost all financially driven endeavors, wealth is proportional to risk.

There are three basic ways to make money from inventing:

1. Invent for a larger entity as an employee receiving a salary.
2. Sell the rights to an invention.
3. Start your own company based on an invention.

Let's look at each of these, as well as how to assert ownership of your invention.

INVENTING AS AN EMPLOYEE

This first approach is the safest, but it results in the least potential financial reward. Large companies all over the world house research programs in which talented individuals are allowed to pursue and develop their own ideas (within the company's overall mission). These invention incubators are often insulated from the short-term, goal-directed culture of the business. Inventors, designers, or researchers (the names vary from company to company) are paid a salary and given tremendous freedom and latitude to create. Organizations such as AT&T Bell Laboratories and IBM's Watson Research Center used this model to great success, especially prior to the government's breakup of these companies as monopolies. Xerox's Palo Alto Research Center (PARC) in Palo Alto, California, is another example that vividly illustrates both the benefits and the drawbacks the inventor faces when employed by a large corporation.

Xerox PARC was founded in 1970 with the intention of creating Xerox's next office automation technology. This new venture could be described as a building that housed a bunch of PhDs with a budget. It was not expected to contribute directly to profits but rather to lead Xerox into future post-copier technologies. By 1970, Xerox had built a tremendously successful business based on its copying technology, and it had the financial wherewithal to fund such an enterprise. It was purposely located far from the corporate center of Xerox in order to foster independent ideas and thinking that would be outside the Xerox corporate paradigm. While the engineers and scientists at Xerox PARC were extremely successful in their endeavors, the business culture at Xerox was too focused on photocopiers to understand and exploit the new inventions that were developed. A clash of cultures resulted, and many of the most innovative people at Xerox PARC either left or were fired. While the freedom and funding they were given created a wonderful environment in which to invent, Xerox's inability to integrate radical new inventions into their business led to great frustration among the inventors. Ironically, Xerox PARC succeeded in creating what became the personal computer revolution. However, it was so far ahead of its time that the parent company could neither appreciate it nor take advantage of it in the marketplace.

Working in an environment where you have the freedom to imagine and create has a paradox built in. You are paid to do what you love, but you are not the master of your creation. Your creation might have the potential to become a successful product and improve life for many people, but those who pay your salary might not share your vision. This is the tradeoff. Not only do you not control whether your invention will make it to the marketplace, but all rights to that invention belong to the company that pays your salary.

I worked at AT&T Bell Laboratories for several years, where engineers were rewarded with patents for their accomplishments. The company's large legal staff would pursue patent applications for good ideas. Bell Labs prided itself on the number of patents received annually—this was the organization's metric of inventive creativity. For the individual engineer or inventor, the patent was like a scout's merit badge. It was also a mark of prestige to hold a number of patents. The process of obtaining and enforcing a patent can be very expensive, especially if litigation is involved. All these expenses were picked up by the company. In return, we had to sign over all patent rights to the company, which then became the owner of the patent. However, the engineer's name was still listed on the patent as the inventor. Most companies operate in a similar manner.

Although an inventor will not get rich on a corporate salary, the corporation provides a relatively safe cocoon from which to practice the art of invention. Funding is available; tools, materials, and facilities are provided; and work takes place within a community of like-minded individuals. Some creative people can find great satisfaction within such an environment. As an employee, the inventor can also learn business and management skills. The road to riches in the corporate world runs through management and business, rather than technical proficiency, and some individuals find those areas more to their liking.

Inventing under the auspices of a large company carries the risk that funding can be sensitive to the financial performance of the organization. Typically, when large corporations need to trim their budgets, the first thing to be cut is research—especially research that is not directly contributing to the bottom line. In recent years, corporations have been much less hesitant to "downsize," which is a euphemism for firing or laying off employees. Inventors are especially vulnerable to this cyclic hiring and firing.

Finally, working for a large corporation can teach an inventor the business skills that he will need before setting out on his own. This can be a good and safe way to learn what is required to make an invention successful in the marketplace. Many entrepreneurs were initially schooled in the corporate world before starting their own companies.

Selling Your Inventions

Next on the risk–reward spectrum is selling your inventions to a third party. This can be done in two ways. You can sell your idea to a company or individual who is in a position to commercialize it, or you can sell your inventive skills on a consulting basis. A "consulting inventor" is akin to a consulting engineering firm selling design-and-build services. In fact, the line between invention and engineering is often very fuzzy. Engineering firms are frequently hired to solve a problem for which a solution needs to be invented from scratch. Inventor-consultants can work as employees for large consulting firms such as McKinsey or Bain, or for smaller engineering design firms. Some are self-employed and work on a project for a set fee. Large consulting firms often reward their employees with significant bonuses in addition to salary based on the company's overall financial success. However, all inventive activity is limited to the problems presented by the client.

Most inventors prefer to sell their inventions rather than their skills. They would like to see their inventions widely used and gain a profit from them without having to start a new business themselves. This is a difficult route to pursue. First, companies that are capable and interested in marketing the invention must be identified. Then the inventor must attract their interest. Many companies are prejudiced against paying for an outside invention. This is referred to as the "not invented here" syndrome, meaning, "If we didn't invent it, it couldn't be any good, and we won't support it." Many companies simply will not talk to an outside inventor. In some cases, this policy is a way of avoiding accusations of stealing an inventor's idea if the company independently develops a similar invention.

Then there are the terms of disclosure. The inventor is faced with how to protect his idea from being stolen once he discloses it. Typically, this is done through a legal document called a nondisclosure agreement, which, in principle, protects the inventor against unauthorized use or disclosure of the idea. However, such agreements can only be enforced through legal action, something a larger corporation can afford more easily than the individual inventor.

While the "not invented here" attitude is still prevalent among many companies, there are some that are open to outside problem solving. Such forward-thinking companies have come to realize that there are many qualified people, outside their own laboratories, who might be able to come up with creative solutions to challenging technical problems. Now, with the advent of the World Wide Web, these companies can reach out to inventors and technical experts all over the world. Web sites such as InnoCentive (www2.innocentive.com) post technical problems from participating companies with an offer of a cash reward for the solution. While this approach can be fruitful in some cases, inventors are limited to trying to solve other people's difficult problems, rather than coming up with their own inventions.

So how does the lone inventor sell his great idea? Selling an idea is similar to finding a partner for marriage. You (hopefully) don't choose at random. And even if you do, most choices might not be interested in you. The best partnerships—and the relationship between the inventor and the buyer of the invention often become partnerships—are those in which each side has a unique need that the other can fulfill.

A beautiful example of this is the birth of xerography. Haloid Corporation, a small company in Rochester, New York, that produced photographic paper, was looking at a troubled future in the late 1940s. Their market share was drastically threatened by several new competitors who were manufacturing photosensitive paper. The company realized that the business needed to branch out into new areas or face a rapid decline.

At the time, an inventor named Chester Carlson was trying to get someone to purchase the rights to his invention of electrophotography, today better known as xerography. He demonstrated his invention to twenty-one major companies, from IBM to RCA to General Electric. None of them saw any potential in the process. Finally, Battelle Memorial Institute, a privately endowed research organization, agreed to fund a continuation of Carlson's research to the tune of $3,000. They asked Carlson to look for a corporate cosponsor.

As the story goes, John Dessauer, the former head of research and engineering at Haloid, was reading everything he could related to photography in search of a new avenue for the company. He chanced upon an abstract of an article written about Carlson's electrophotography process. Although he felt he was grasping at straws, Dessauer decided he had nothing to lose by visiting Battelle and seeing what electrophotography was all about.

The "marriage" began with a first date, as the Haloid executives traveled to Battelle to investigate this new process. Although they felt the process was far from commercialization, they decided that electrophotography had potential and agreed to be copartners with Carlson and Battelle in continuing to develop the process.

The story is detailed in Dessauer's book *My Years with Xerox: The Billions Nobody Wanted*,[1] and we all know the eventual outcome. What most of us are not aware of is how far electrophotography was from commercialization, and the tremendous risks that Haloid took in funding and pursuing its development. The story is a fascinating example of tenacity and vision on the part of a small company and a group of committed executives.

The Xerox story shows that it is possible to enjoy tremendous success from selling an invention. However, both parties must feel that they have something to gain from the relationship. Haloid was desperately looking for new products as competition threatened its core business. Chester Carlson was just as desperately seeking a partner who could make his idea into a commercial reality. Through the initial auspices of the Battelle Institute, the match between the two was made.

Unfortunately, the Xerox story is more the exception than the rule. The lone inventor is usually at a great disadvantage in trying to interest a manufacturer or marketer in paying for an invention. If, however, one succeeds in selling an invention, payment schemes can range from a lump sum payment to a royalty on sales. Lump sum payments are usually easier for a company to swallow than a continuing royalty whereby the company's obligation extends into the future. However, the advantage of a royalty payment for a company is that it mitigates its risk, because royalties are usually based on sales.

Walking into a company with an idea or even a prototype to sell is a difficult proposition, but companies become more receptive if there is proven value behind the invention. One form of proven value is a patent on the invention. A company will need to be convinced that the patent is substantial and capable of defense against legal challenges by competitors. If a company finds the invention interesting, a patent adds value by ensuring exclusive rights to the invention.

An invention will also attract more commercial interest if it has already been successful in the marketplace and has begun to produce a revenue stream. By the time your invention has reached this stage, you may also have established a brand name and brand recognition. Now you are speaking the language of business, and companies can more easily evaluate what they will be purchasing from you. Of course, you are an inventor and not necessarily a businessperson. Inventing and marketing are two very different disciplines. However, as we will see in the next section, the two can come together, as the inventor strives to get his invention to see the light of day and to profit from it.

Entrepreneurship: Starting Your Own Company

What if the inventor does not want to sell his invention to someone else but instead wants to use it to launch a business venture of his own? This is the riskiest of the three strategies, but it can also be the most financially rewarding. However, it requires a new set of skills: the skills of entrepreneurship.

An entrepreneur is someone who starts his own business—usually driven by a novel idea, invention, or opportunity that he feels can be turned into a profitable enterprise. For the inventor, starting one's own business is venturing into a completely new area of endeavor. There can be great financial risk, and the demands of managing a business can easily take the inventor away from the creative process of inventing. Many inventors have found the path of entrepreneurship to be very gratifying; others have found themselves to be completely at a loss in the business world.

Every business is unique, but at the most basic level, an operating business needs to be able to do the following things:

- make its product in sufficient quantity and quality
- promote and sell the product to its customers
- deliver the product to the customer in a timely manner
- manage cash flow and other financial and administrative matters

Let's start with manufacturing.

Manufacturing

In the past, most companies manufactured their products in their own facilities. Now many companies rely on contract manufacturers—businesses that specialize in producing a product to a specification. Contract manufacturers might charge a flat fee, a percentage of future sales, or both. Companies that use contract manufacturers may perform only the final assembly of their product or they may simply market and distribute the product. Using a contract manufacturer is the easiest way for you to turn your invention into a mass-produced product. In order to reduce the cost of manufacturing, there will probably need to be changes in both the design and materials of your prototype. As the inventor and design engineer, you will negotiate your way through this process, deciding what trade-offs you will make to reduce the cost and improve the efficiency of manufacturing.

The technical aspects of dealing with a contract manufacturer are usually straightforward, as information and drawings can easily be exchanged online. However, the business aspects can be more challenging. The contract manufacturer is motivated by the quantity of product it will manufacture. As you launch your business, you might not be prepared to order ten thousand or one hundred thousand units. This will be a key point of negotiation. How much risk are you willing to take? One approach is to negotiate a ramp-up scheme, in which you agree to buy a large number of units in the future in exchange for a manufacturer's willingness to start production with a relatively small initial order. However, contract manufacturing always involves the risk of paying for goods in advance on the assumption that you can sell them.

Selling

If you can't sell your invention, all other efforts are for naught. The first part of the selling process is to establish whether people are interested in buying the product and what price they are willing to pay. Established businesses conduct detailed market studies to try to determine this information. With completely new products or new markets, these studies often turn out to be little more than educated guesswork. The entrepreneur, lacking the resources of an established company, often relies on gut feeling or intuition. This "feel" for the marketplace is, in my opinion, the key to a successful entrepreneur. If everything could be quantified and calculated, then computers would be running businesses, not humans. Still, the risk is yours, and it is worthwhile to gather all the information you can on the salability of your invention. Will enough people be willing to pay enough money for your product for you to make a profit? You don't want to have great sales but lose money on each item sold. The old business joke of selling at a loss but making it up on volume will fall pretty flat for one who finds himself in that position.

Once you determine that you will be able to market your product profitably, you will need to establish a means of selling, delivering, and receiving payment. In the past, this would have involved either contracting with a distributor—a company that warehouses your product and employs salespeople to actively promote and sell it—or incurring the cost of selling the product yourself. Today, the Internet provides a much less expensive way to reach the marketplace. You can set up a Web site from which you can advertise and sell your product. To make it even easier, you can link directly from your site to companies that will warehouse and ship products to customers and manage your financial transactions. The beauty of the Internet is that you can focus on your new ideas and pay others to perform the routine tasks that make your business function. The Internet gives the entrepreneur an instant reach into the marketplace where, with little up-front investment, he can test, refine, and grow his embryonic business. It allows the entrepreneur to bypass the traditional sales channels that require greater initial sales volumes and investment.

So, conceptually, an ideal arrangement for selling your invention might consist of a Web site that promotes and sells the product in conjunction with agreements with various vendors to make, store, and ship the product as well as manage payment. In this ideal world, you could, hypothetically, have a hands-off business where all you have to do is create the invention, and everything else is subcontracted.

Administration and Logistics

I've made the business side of things seem easy so far, but it really isn't that simple. The model described above is an ideal model—the way you would like things to work. Murphy's law holds in business as it does in other fields. I have outlined the basic plan for a business without mentioning the many pitfalls, and maybe that is the right thing to do. Many successful entrepreneurs have said that if they had known at the beginning how hard it is to create a successful business, they never would have attempted it.

In the medical field, the routine, boring, detailed work is referred to as "scut work." The truth is that scut work is the difference between success and failure. Most entrepreneurs spend an inordinate amount of time dealing with the scut—the frustrating details and ridiculous problems of their incipient businesses. Administering the detailed workings of your business can take up most of your time. If you were to do everything yourself, everything would be done just the way you want it, but, of course, that is impossible. In the scenario discussed above, I have included contractors—third-party specialists who do all the things that you don't want to do and who are supposed to make your life easier. However, these specialists have many other clients besides you. You still have to keep track of what they do and make sure everything is being done to your liking. For example, how will you monitor product quality if you never see the product? How will you manage inventory, production forecasts, and backorders? These are only a few of the manufacturing issues that need to be addressed. How will you ensure timely delivery and manage cash flow? What information systems will you need to keep track of everything? As your business grows, there are many administrative systems that need to be set up in order to ensure that everything functions according to plan. Success leads to new problems that you never dreamed of when you invented your wonder product. This is one of the reasons that many inventors sell their businesses to other companies once they become established and start to grow. The operation of a business is a specialty within itself and not necessarily within the purview or skill set of the typical inventor.

Finance and Cash Flow

The entrepreneur is like a juggler, but instead of juggling balls or bowling pins, he juggles cash. Every entrepreneur needs cash to start and run his business. Some people use their own savings as funding to start their businesses; others borrow. Venture capital firms provide money to new or promising enterprises in return for part ownership of the business. Many inventors, however, prefer to retain control of their businesses. They are more likely to depend on their own resources or bank loans.

Managing cash flow can be one of the major challenges for the entrepreneur. He realizes that he has a great product and that people will line up to purchase it, but he also knows he has to put up a good deal of cash to get it manufactured, shipped, and sold. The time lag between cash coming in and going out needs to be closely managed if the potentially successful entrepreneur is to avoid ruining his business and incurring large personal debts. A loan can help to span this gap, but the debt adds additional cost in the form of interest payments. Many entrepreneurs have dug themselves into financial pits from which they cannot climb out.

The phrase "the bottom line" comes from the last line at the bottom of a financial statement, which describes the final profit. In the end, a business needs to have enough cash at all times to run its operations and eventually make a profit. This is the bottom line for a sound business. While there are emergency levers that can be pulled, such as borrowing money, if the business cannot eventually succeed financially on its own, it is doomed to failure, no matter how great its products.

Starting a business is not that difficult. Running it and making it a success is. I have only briefly touched on what is necessary to start and run a successful business. Typically, entrepreneurs learn as they go along. One of the biggest stressors for a business is growth. Achieving stability is often not an option in a competitive environment. As soon as you think you have mastered the challenges of profitably making and selling your product, a competitor pops up with a similar but less expensive product, and you need to rethink your entire strategy.

All this said, entrepreneurship offers tremendous benefits. For one, you can reap the maximum financial reward for your invention. You can also build a single invention into an entire product line and business. The risk that a small company called Haloid took on an unproven process, which the lone inventor called electrophotography, spawned Xerox Corporation. But we must not overlook the perseverance of the Haloid executives over years of frustrations with the embryonic technology of dry copying. Haloid's initial investment in the process was made in 1946. The first commercial copier, the Xerox 914, was not marketed until 1960.

While most entrepreneurs would have neither the financial resources nor the patience of the executives at Haloid, if their product is right for the market, they can establish a small business in a relatively short time. As the business and its challenges grow, they can look to sell at any time. Once they have shown success by producing a solid stream of revenue, others—often larger companies—will be more amenable to purchasing the business. Typically, purchasers are motivated by a proven product, business, or brand that they believe has not been exploited to its fullest potential.

Entrepreneurship is not the path for all inventors. It is fraught with risk and requires business skills and intuition that the inventor might not have. It also requires a more aggressive, extroverted attitude that might be uncomfortable for a more introverted inventor. In addition, there can be great financial risk involved (unless one has a venture capital firm underwrite the business in return for full or partial ownership). With all this taken into consideration, entrepreneurship can also be the greatest opportunity for an inventor. Once a business is established, the inventor has a platform for selling future inventions. Once a brand name is established in the marketplace, consumers will look for more products marketed under that brand.

I would like to conclude with a personal example. My father, Edward Paley, who was an entrepreneur before the word became fashionable, started a small company in the basement of our house in 1963. The company was based on a single product, a textile that he invented for cleaning electromechanical components and optics that needed to be used in environments where there could be no outside contamination. He continued to invent and market new products around the theme of "contamination control for critical environments." The business, which he financed through a loan from his father-in-law, was initially tenuous. Fortunately, the only employees were his wife and children (we didn't have to worry about downsizing). However, the market need was real, and the business served as a platform for his future inventions. The business, named Texwipe after its first product, grew and produced new challenges with each growth spurt. With the advent of the semiconductor industry in the early 1970s, the business took off and grew to become a multinational corporation and a leader in the area of contamination control.

I tell this story not only out of personal pride but as a true-life example of an inventor becoming a successful entrepreneur. There are many others in the world like my father who were able to use their vision and creative talent to surmount the challenges of business and add innovation and vitality to the marketplace.

PATENTS

I would be remiss if I did not address the issue of patents. Even though patents are often thought of as an indicator of inventiveness, I believe that patents are tied exclusively to the commercialization of ideas. Patents give legal rights to an inventor, making the inventor or an assignee the owner of the intellectual property the invention represents. A patent is granted not necessarily on the merit of an idea but on whether an idea can be shown to meet three legally defined criteria: novelty, non-obviousness, and utility. Patenting an invention gives the inventor or assignee the proprietary right to sell the invention while excluding others from doing the same. A patent is divided into two sections: an overall description of the invention and detailed claims that specify the invention. It is the claims that legally define the metes and bounds of the invention. The Patent Office, in reviewing the proposed claims, may accept them, reject them, or require that they be modified. Typically, the inventor will want his claims to be as broad as possible.

Businesses use patents extensively to create exclusive rights in a specific market area. The pharmaceutical industry business model, for example, relies on patents to create a monopoly for a new drug for the twenty-year period of the life of the patent. This enables the company to recover the large research and development costs associated with the creation of a new drug.

However, patents have a downside. When filing a patent application, the invention must be sufficiently disclosed so that one "skilled in the art" can duplicate it. Once the patent is granted, it becomes public information. Potential competitors will then be able to replicate it and try to design a competitive product that skirts around the claims listed in the patent. Often, patent violation is a nebulous area that is resolved only after an extensive and expensive court battle.

The individual inventor can be at a disadvantage when trying to prosecute a patent infringer. If the infringer is a large corporation, money rather than merit might well decide the case. One should look at patents as a business tool rather than a mark of inventive competency. The expense of the application process and possible patent defense—not to mention the time and aggravation involved—place patents more in the purview of a corporate legal department than that of the individual inventor. Even for large corporations with extensive resources, the question of whether to patent an invention is complex. Most patent attorneys advise weighing the risk of disclosure against the breadth of your possible claims. In other words, once the invention is disclosed, how easy will it be for someone to produce something similar that navigates around the patent claims without infringing on the patent? My feeling in this regard is that trade secrets in industry can be as valuable as patents. I would say that if the invention is a product for all to see and use, then it should be protected by a patent. Once such a product is out in the marketplace, it is easy for a competing company to reverse-engineer it and make a copy. Patent protection can help in this case, as long as you are willing to spend the money to defend the patent in court. On the other hand, if the product is produced by a unique process, and that process is the invention, it makes more sense to keep the process proprietary and not disclose it by filing a patent. These are business decisions that are based on financial realities and objectives.

There have been cases in which the lone inventor, armed with a patent on his invention, has triumphed. As you can imagine, it is very difficult for an individual to go into a legal battle with a large corporation and win. One of the most famous cases is the invention of the intermittent windshield wiper by Robert Kearns.[2] Kearns sued Ford Motor Company for infringing on his patent, which Ford claimed was invalid. Kearns became obsessed with his cause and pursued infringement suits against Ford for twelve years and against Chrysler for thirteen years. There are very few inventors who have both the financial and the emotional wherewithal to fight such battles. In the end, the jury awarded him $10.2 million from Ford and $18.7 million from Chrysler. Such victories, however, are by far the exception rather than the rule.

Obtaining a Patent

The process of applying for and obtaining a patent can be relatively short and painless or long, expensive, and tortuous. It all depends on how the patent claims you submit are viewed by the patent examiner. To describe the process in great detail is beyond the scope of this book. I strongly suggest, especially for those who are unfamiliar with the process, that you engage the services of a competent patent attorney. If you are considering applying for a patent in the future, you should date and document each important event in the conception and development of your invention. This will be important if your patent is disputed and you need to prove that you invented the product or came up with the seminal ideas before anyone else. If you have been documenting your invention along the way with written descriptions, drawings, or prototypes, you should have them legally witnessed periodically to establish a time and date. Typically, scientists who work for large corporations are required to keep bound notebooks in which they document their work. In many cases, a notary public will sign a particularly significant page or group of pages, witnessing the date and time. For the individual inventor, it is also a good idea to get key drawings or descriptions witnessed. This can be easily done by a notary public at any government office or public library.

Once you engage the services of a patent attorney, the first thing the attorney will do is review your idea against prior art. *Prior art* includes anything in the public knowledge, be it literature, products, or public descriptions of any kind that could be construed to describe the essence of your idea prior to your recording of the idea. On the basis of this preliminary research, the attorney will provide you with an opinion as to whether the idea is novel, nonobvious, and can be deemed useful. This initial research will cost roughly $1,000 to $3,000 and provides you with an expert opinion as to whether your idea is likely to meet the criteria for a patent.[3]

If you decide to apply for a patent based on a preliminary review by your attorney, he or she will write up a patent application with your guidance. The application consists of a description of the idea and specific inventive claims that legally define the scope of your invention. The attorney will then submit the application to the Patent Office for review. The cost of writing and submitting an application can range from about $5,000 to more than $15,000. These costs can vary greatly depending on the attorney, geographical area, and the complexity of the invention.

Patent attorney Gene Quinn has developed the table shown in figure 9.1 as a rough guide for using the complexity of an invention to estimate the cost of preparing and filing a patent.[4] "Complexity" in this context includes both the physical structure of the invention and the effort involved in preparing the patent application.

Complexity and cost of patent preparation

Type of Invention	Examples	Cost
Relatively Simple	electric switch; coat hanger; paper clip; diapers; earmuffs; ice cube tray	$4,500 to $6,000
Minimal Complexity	board game; umbrella; retractable dog leash; belt clip for cell phone; toothbrush; flashlight	$6,000 to $8,000
Moderately Complex	power hand tool; lawn mower; camera; cell phone; microwave oven	$8,000 to $10,000
Intermediate Complexity	ride-on lawn mower; video game; simple RFID devices; solar concentrator	$10,000 to $12,000
Relatively Complex	shock-absorbing prosthetic device; Internet implemented business method with computer system	$12,000 to $15,000
Highly Complex	MRI scanner; PCR; telecommunication networking systems; software	$15,000 +

Figure 9.1: Patent complexity and cost table.
Courtesy of patent attorney Gene Quinn of IPWatchdog.com.

Depending on the nature of the invention, your application can sit in the Patent Office for a period of several months to over two years before it is reviewed by an examiner. The examiner will then render an "office action," which will be mailed to the attorney. The office action will be a recommendation to either allow the patent, reject the patent, or ask for specific modifications. Typically, patents are rejected the first time around for specific reasons, such as excessively broad claims.

If the patent is rejected, you and your attorney will review the reasons for the rejection and amend the patent application. The examiner will consider the amended application (usually in a shorter time period) to see whether your changes have resolved his objections. While you are trying to write your patent so that it is as broad as possible, the examiner is trying to make sure that it is as specific as possible and does not overstep into any prior art. This process of going back and forth with the Patent Office, called "prosecution," can go on for many cycles, sometimes taking a number of years. If there seems to be no progress, it is often worthwhile for you and your attorney to meet with the examiner to try to reach a mutually agreeable compromise.

Eventually, the patent examiner will either allow or reject your application. If the application is allowed, the Patent Office will issue a patent. If the application is rejected, you can still appeal the rejection to a higher authority within the Patent Office. There, the application will be reviewed again and a judgment rendered.

The cost to finally obtain a patent can add an additional $1,000 to $5,000 to the preparation fee. These costs include additional attorney fees for prosecution, if necessary, and patent issue fees.

Securing a patent can be expensive and time consuming. In the best-case scenario, the process takes about two years and costs around $5,000. If the patent is very complex or requires multiple submissions, the cost can climb to well over $20,000, and the process can take up to five years or longer.

Once you have the patent in hand, you have the right to exclude others from making, using, or selling the invention in the country in which the patent has been granted. However, as discussed previously, it is up to you to uphold that right. If you believe that someone is violating your patent, you need to use the legal system to confront and stop them. If it comes to this point, the other party can (and probably will) claim that your patent is invalid (even though it was granted by the Patent Office!). They will bring examples of prior art that perhaps were never considered or look for legal flaws in your application. You can easily find yourself in the position of defending the validity of your patent in court. Although the Patent Office gave you its stamp of approval in issuing the patent, the patent is only as good as your ability to defend it in court.

It is possible to save on attorney fees by being your own patent attorney and doing everything yourself. Some inventors feel comfortable doing this, especially those who have gone through the patent process many times. However, for the novice, it is best to seek the advice of an expert and to pay for an experienced patent attorney to pave the way.

Patents are a very complex subject, and I have provided here only a brief introduction to the topic. The most important fact to remember about patents is that they are a commercial tool for obtaining a time-limited exclusive right to make, use, or sell an invention. Although many inventors look at patents as "badges of honor," the reality is that they are business tools that are useful in commercializing an invention. The decision to seek a patent needs to be evaluated in that light.

The Business of Invention

Most inventors don't become rich. They invent for the same reason that artists paint or writers write—to fulfill their own creative impulses. That said, one can still profit financially from inventing. In this chapter, I have described three established ways to do this. There are other, more peripheral ways as well. Teaching invention in an engineering school or consulting on the creative or problem-solving process for corporations are two other possible ways to gain remuneration as an inventor. An inventor's creative talents can lead in many different directions. However, the passion for invention, which is what drives most of us, will inevitably bring us back to the workbench, where we will dream up and build our new ideas. There is an aphorism attributed to Ralph Waldo Emerson and used in many 1950s college graduation speeches that decrees: "Build a better mousetrap and the world will beat a path to your door." If only this were true. A more accurate claim would be that once you build a better mousetrap, it is up to you to find a way, whether through employment, selling it, or starting your own business, to get it out into the world so that both you and the world can reap the benefits.

Chapter 10

THE ART OF INVENTION

"Inventing is not a logical process. It's only logical after the fact."
—Jacob Rabinow

Inventing is an art. The creative process defies mechanistic explanation. It is easy to talk around the periphery of how to invent, but it is impossible to give a cookbook recipe for invention. In my opinion, this is what makes inventing so exciting. There are always surprises and sudden insights that pop out and fill you with wonder. Did I really come up with that idea or solution? Of course you did. Could you have envisioned it from the start? Definitely not. In his biography of Albert Einstein, author Walter Isaacson quotes from a letter Einstein wrote in 1949: "A new idea comes suddenly and in a rather intuitive way. But, intuition is nothing but the outcome of earlier intellectual experience."[1] The more deeply the inventor delves into his subject area, the more material there is for the creative brain to process. The creative process breaks observations down into their fundamental components and reconstructs them in a new way. The new way offers a solution or insight into a problem that could not have been imagined previously.

THE UNEXPLORED: HARD WORK, FEAR, AND FANTASY

Invention at its most fundamental is making things up. Children naturally tend to imagine and create through fantasy. While this is considered normal for young children, as we grow older, this trait is discouraged. We are not rewarded in school for daydreaming. However, as inventors, we must regain this childlike ability. We must not be afraid of making things up. After all, something new does not necessarily have a clear past to build on.

The art of invention is a combination of many things. It involves a deep understanding of the interplay between the conscious and the unconscious mind, it requires the patience to let this interplay run its course, it requires the intuition and insight to manage and manipulate the flow and formation of ideas, and it requires hard work—delving deep into the subject matter surrounding your area of invention. I referred to this previously as feeding the brain. If, as Albert Einstein put it, "[i]ntuition is the outcome of earlier intellectual experience," you need to make sure that you are fully enmeshed with your subject matter. You should wear your subject area like clothing. It should become an obsession. This is the necessary preparation for the creative breakthrough.

Finally, and perhaps most important, the inventor must have courage. Unless you have the courage to go forth and create, your great ideas will remain locked in your imagination. The inventor is like an explorer stepping off into new and unknown territory. There are no landmarks or signposts to show the way. You must have the courage to overcome any nagging self-doubt and push forward with your new idea in the absence of outside reassurance or validation. Courage and belief in yourself are essential to propel you forward into the unknown territory of invention.

Experience and Invention

In chapter 4, we examined the influence of life experience on invention. This influence cannot be overstated. After all, our experiences make us who we are. Our life narratives are told through our experiences. I think that our creative acts are also a product of experience. For example, Einstein explained his theory of special relativity in terms of clocks and trains. He performed his thought experiments and derived the theory while working as a patent examiner in Bern, Switzerland, in 1905. The town of Bern had a large train station with rows of clocks synchronized to a main clock tower in the center of town. Einstein walked past this tower and train station every day on his way to work. His job as a patent examiner required him to review electromechanical devices, including a myriad of applications on ways to synchronize clocks set apart from one another. Now, not everyone who gazes at a train station or works with clock patents comes up with the theory of relativity. But it has often been suggested that these experiences gave Albert Einstein the physical vocabulary with which to describe his revolutionary theory. In other words, he used elements from his real-world experience to embody the thoughts that he was constructing in his mind.

Sometimes seemingly random or unrelated experiences play a significant part in a creative breakthrough. Steve Jobs, the chairman of Apple Computer, told a story to Stanford graduates at their 2005 commencement about how he came up with the idea of using typography and multiple fonts with computers. Jobs dropped out after his first semester at Reed College, though he remained at the school, sleeping on the floors of friends' dorm rooms and making ends meet by returning empty bottles for the deposit. One of the advantages of not being a student enrolled for a degree was that he had no course requirements. So, simply out of curiosity, he sat in on a calligraphy course. In that course he learned all about typefaces, fonts, letter spacing, serif versus sans serif, and what it took to make the printing of letters a beautiful art form. The course had no practical value that he could see, but he took pleasure in gaining an appreciation for typography. Ten years and an entire world later, Jobs found himself involved in the brand-new field of personal computing. The seemingly random experience of having sat in on the typography course at Reed College became an impetus for the design of the Macintosh computer user interface and set a standard that emphasized typography and font choice for all word processing software to come.

Jobs's message to the graduates that commencement day was to follow their instincts and curiosity, to soak up all the experience available, for one never knows when or how that experience will come in handy. There is an important message for the inventor here. Never turn down the opportunity to learn something. Everything you learn or experience can influence your inventive thinking.

The Art of Getting It Wrong

When we search for the new, there is a chance we might not find it. We might be wrong in our assumptions, we might be mistaken in our direction, we might be naive in our goals. In short, we might make mistakes. Mistakes go against our cultural grain. Isn't it better to be right than wrong? How many teachers have congratulated students for coming up with the *wrong* answer? We are taught from early on that *right* is good and *wrong* is bad. Morally, that may be correct, but for inventors such an attitude is the kiss of death. The fear of being wrong militates against the willingness to take risks. Without risk, there is no invention. Without risk, there is nothing new. The nature of the art of invention is that you will make many mistakes. You will go out into unexplored areas that you have to navigate purely by intuition and feel, and you will not always choose the right path. One of the great challenges of inventing is becoming comfortable with being wrong. Your mistakes will often guide you to the correct answer, if you take the time to analyze them. Mistakes are hardly ever wasted effort, for they serve as a form of instruction. Sometimes their lessons come to light only at a later time, but mistakes can be our greatest teachers. If we are not willing to make mistakes, we pass up tremendous opportunities to learn. By the same token, we must learn to recognize when we are wrong. Once we realize that we are headed down the wrong path, we should be able to step back and regroup. We should never let our hubris propel us in a direction that we know is wrong, simply because we are emotionally invested in it.

Although we learn from mistakes, it is still uncomfortable to be wrong. You, the great inventor, have invested much time and effort in a direction that you now realize is incorrect. What do you have to show for it? You have the newfound knowledge that the path you have taken is not the right one, but you have nothing of substance to show for this knowledge. If you are asked for a progress report, all you can say is what shouldn't be done. This is a very uncomfortable feeling but is nevertheless an important part of invention. What doesn't work will often lead you to what does.

Think of the mathematician working on a complex proof. He attacks it from many angles, creating pages of equations. Time after time, his efforts come to naught. He throws the pages of work into a discard pile. Then one day, he succeeds in cracking the problem; he discovers the correct approach and formulates just the right equations to satisfy all necessary conditions. Perhaps he has worked weeks, months, or years on this problem. Could he have found the right path without traveling through all the wrong ones? It is highly unlikely. Perhaps he could have made certain connections earlier than he did or realized the importance of some initial thoughts that he abandoned prematurely, but the entire process of discovery and creation requires an effort that is almost always filled with mistakes. Even sudden flashes of insight are often couched in previous mistakes.

The cliché that one needs to learn from mistakes is often voiced as a balm for having made them. As an inventor, making mistakes is part of your job. It is not something to be embarrassed about or something that needs to be soothed away. If you aren't making mistakes, you aren't trying hard enough; you aren't stretching or pushing yourself. Getting it wrong after a serious effort shows that you are doing something right. True novelty in invention is often built on a foundation of errors.

The Hidden Obvious Revisited

In chapter 2, we spoke about seeing the hidden obvious. How is it that we are unable to see something that is in plain sight? The answer is that we see, but we are unaware. The art of invention requires a heightened awareness—an almost unnatural awareness of the potential of everything around us. One of the biggest obstacles an inventor faces is the mind's filtering and categorization of information. Conditions such as schizophrenia and autism are thought to result from the inability to filter and categorize information. These are conditions of hyperawareness resulting in mental overload. To lead a functional life, we need to filter out most of the information that we sense. The cost of doing this is that we desensitize our awareness. When we take in information from our environment, we immediately want to make sense of it. We want to place it in a category. The problem with this natural (and necessary) tendency is that we sacrifice the ability to see things in a new way. Perhaps the overriding trait needed to overcome dulled awareness is curiosity. Intense curiosity will always keep you on the lookout for the hidden obvious.

Curiosity leads us down paths we might otherwise never take. Curiosity means pursuing something that does not necessarily have an obvious purpose. We examine something and follow the path that we are led along merely because it seems interesting. Earlier, I quoted the well-known Robert Frost poem "The Road Not Taken," in which Frost talks about "how way leads on to way." This is a natural consequence of curiosity. Steve Jobs was curious about calligraphy and decided to sit in on a class, most likely without any obvious goal in mind. Curiosity leads to connections—often seemingly bizarre connections that can form the basis for new ideas. In Jobs's case, his curiosity eventually led to the use of multiple fonts in word processors, something that is now taken for granted. One could even say that such an invention is obvious. But take yourself back to the days of typewriters, when computers were scientific and business machines, and the primary mode of documentation was typing. Most people did not even know what the word *font* meant, and if they did, they would be hard-pressed to name a single one. What kind of thinker would have blended the idea of multiple fonts, or calligraphy, with what were then "calculating machines"?

Most great inventions are obvious only in retrospect. Many of them seem absolutely crazy when they are conceived. The hidden obvious is almost always present, but it takes the curious thinker, unfettered by convention, to discover it.

SERENDIPITY

A not-so-secret secret is that sometimes inventors just get lucky. Sometimes we are favored by chance. The dictionary definition of the word *serendipity* is "an aptitude for making valuable discoveries by accident." This seems like a contradiction. How can one have an aptitude for making something by accident? As strange as it seems, serendipity plays a significant role in inventing. Sometimes, when we work hard enough on a particular project, we unconsciously perceive things that might ordinarily go unnoticed. We say we got lucky, but I don't believe that our luck is completely random.

There is a fascinating story about the chance invention of a blockbuster class of drugs called benzodiazepines that beautifully illustrates this idea. Dr. Leo Sternbach, a native of Austria and Poland, received his PhD in organic chemistry from the University of Krakow. His specialty was textile dyes, more specifically, the formulation of diazo-based dyestuffs whose chemical structure made them ideal for permanent dyes that could be used on fabrics and other materials. As a refugee from the Nazi regime during the Second World War, Sternbach found employment with Hoffman-La Roche in Basel, Switzerland, before being sent by them to Nutley, New Jersey. There, he was initially assigned to work on new antibiotics. By the 1950s, the strategic focus of many pharmaceutical companies had expanded to include the development of psychiatric medications. In 1955, meprobamate, known popularly as Miltown, was marketed as the first minor tranquilizer. The immediate success of meprobamate led many pharmaceutical companies to seek their own compounds. Roche understood the commercial importance of finding such a tranquilizer, and Sternbach was summarily pulled from his work on antibiotics and given this challenging assignment.

The starting point in this search was the meprobamate molecule, and Sternbach explored all possible variations without success. He wondered to himself whether there was another place to begin. Certainly this was completely different from the work he had been doing previously. He could not synthesize a new tranquilizer from an antibiotic. He thought about it and decided to go back to his original field of expertise, diazo dyestuffs, as a chemical starting place. The truth is, he could have started anywhere, but it is a truism that when one tackles an unknown problem, one often returns to previous knowledge as a firm place from which to start. Sternbach was no different. Since he had much experience with the chemistry of diazo dyes, and he thought they could possibly be biologically active, why not start there?

He went to work, and for the next two years he synthesized and tested around forty compounds based on an aromatic diazo chemical structure. Nothing worked; they were all pharmacologically inactive. As he searched for more variations to try, he synthesized a chemical based on a diazo ring with a methylamine side chain. He put the new chemical on the shelf for the next round of testing. The next round, however, did not come. Roche looked at two years of failure on the project and decided to "pull the plug." The company felt that it was not even close to achieving its new miracle drug and that further effort would go nowhere. Sternbach was told to drop the project and go back to his work on antibiotics. He cleared off his desk and started on his new assignment.

Eighteen months later, while working on the antibiotic project, his laboratory assistant decided that the lab was a mess, and it was time to clean up and throw away the accumulated jars of old chemicals that were sitting on Sternbach's shelves. As he proceeded to clean up the lab, he ran each chemical past Sternbach before disposing of it. One of the jars he picked up, labeled Ro 5-0690, held the untested compound from the previous project.[2] He showed this to Leo Sternbach, asking whether he should throw it away. Sternbach thought for a moment. As he later related to his friend and colleague Dr. Jonathan Cohen,[3] he thought that the structure of this particular chemical might well have biological activity, and before discarding it, he wanted to test it. Even though he was no longer involved with the project, he asked the assistant to submit it to Dr. Lowell Randall, Roche's head of pharmacology for animal testing.

The results of the animal tests astounded the scientists at Roche. The initial results showed that they had found the tranquilizer they were searching for. After clinical studies were completed, the drug, now called chlordiazepoxide, was released under the brand name Librium in 1960. It was the father of a whole generation of drugs, the best known of which is Valium, based on the same structural chemistry.

Serendipity? Gut feeling? Intuition? All of these, as we have seen, are keys to invention. Did the monumental discovery of benzodiazepine-class drugs lie in a split-second decision about whether to discard an old chemical from a failed project? Over the eighteen months that passed between Sternbach putting the chemical on the shelf and taking another look at it, did his unconscious process the significance of this particular structure (which was unique from his other compounds)? All these possibilities are, of course, speculation. But one thing we do know is that this is a perfect example of how the ephemeral inventive process works. We can never predict when the flash of insight will occur and when serendipity will come into play.

With the success of this chemical, Leo Sternbach was inspired to keep going and see what other drugs he could coax out of the diazo structure. The next one selected for clinical trial was diazepam, better known as Valium, which was released to market in 1963 and became one of the most prescribed pharmaceuticals in the United States for the next twenty years! Valium is still widely used today. Dr. Cohen once asked Sternbach how, from among all his diazo-based experimental compounds following Librium, he chose Valium as the next candidate for clinical development. Sternbach smiled and said he simply thought that this chemical had the best possible profile.

Serendipity, once again.

The Staircase of Creativity

Are we always creative? Is constant creativity a necessity for invention? The answer, of course, is no. No one can be constantly creative. Creativity usually comes in spurts and requires a period of gestation. I refer to this process as the staircase model. If you examine the stairs that you probably climb every day, you will notice that each step has a flat horizontal section called the tread, or run, and a vertical section called the riser. Interestingly, the run is longer than the riser. We seem to need more space and perhaps more time to stabilize ourselves before continuing our ascent. It is the same with invention. There is usually a period of latency that precedes a creative burst—a period where nothing seems to be happening. However, as described earlier, the unconscious mind is working even when nothing seems to be happening. Inventors, like other artists, can become depressed during this period. One's attitude can be characterized by thoughts such as "I am completely blocked" or "I will never be able to think of another good idea again." All creative people go through this, and as with stairs, the run—or latency time—is usually longer than the rise. If our creativity stems from unconscious connections, we need to accept that these connections are not made according to any predetermined schedule.

I clearly remember how depressed I became after what I thought was my biggest inventive triumph in industry. The product was extremely successful, and I felt lousy. "Well, now I'm done," I thought. "There is no way I can top this." The period of what I considered my slump lingered much longer than I would have liked. In fact, it lasted more than a year. I honestly felt that I had reached my inventive peak and that everything was downhill from there. In truth, I had the wrong picture in my mind. Yes, we have peaks of creativity during which inventive ideas flow, but these peaks inevitably recede, and with time, return once again. However, it is very difficult to see this when you are coming off of such a peak. In my case, after a long, latent horizontal "run," my creative ideas burst forth again, and I achieved an even greater success with my next invention. The analogies are endless. I use the staircase as an example, but another picture of the process might be a wave with its peaks and troughs. However you see the process, it is important to accept that constant creativity is not sustainable, and that there will be considerable periods of time during which you will not be able to create. It's not that nothing productive is happening, it's simply that bursts of creativity take time to build before they can be expressed. You can look at this as the creative cycle.

The Pleasure of the Problem

Pleasure and *problem* are two words that are not often associated with each other. The terms *pleasure* and *puzzle* form a more likely association. In truth, we derive a great deal of pleasure from solving problems. We don't usually realize this when we are in the midst of the process, but once the problem is solved, we feel great satisfaction. We intuitively understand that some problems, such as games and puzzles, or more abstract challenges, such as mathematical problems, are indeed pleasurable, and solving an inventive or design problem should also be viewed in this way. There is great satisfaction in working out a problem until a simple, elegant, and robust solution presents itself. This process of working your way through a problem, usually with much iteration in search of an optimal solution, is what makes inventing so satisfying. It is as if you are reeling in a big fish as you alternate between applying tension and giving slack. You have periods of intense focus and then you let go; you create and then you step back. Often, you realize that you must erase what you have created and start over (although your past creation is what gives birth to your new one). As you work, you internalize the problem. Your mind turns it over day and night. You try different solutions, and eventually, things clarify, and the right solution or direction becomes obvious. This experience is somewhat stressful, always exciting, and very creative. It is why we do what we do. We rationalize that the end result justifies the process, but the reality is that we thrive on the process. We actually get pleasure from working through the problem.

The Inventor's Tool Kit

The inventor's tool kit is knowledge and experience. The knowledge, in the area of physical inventions, is an understanding of the fundamental principles of physics, chemistry, biology, and engineering. I'm not talking about schoolbook learning, where you learn what is necessary to pass a test. I'm speaking of a deep, intuitive understanding. This is the only way you will be able to use these principles in practice. For example, can you think of some way that the conservation of angular momentum could be applied to solve a fundamental problem? Or is there something you can take from the mechanism of cell respiration that might be the cornerstone of your next invention? Having a working knowledge of these "tools" readily accessible gives the inventor a store of solutions to draw upon. Intuitive understanding comes from the experience of using an abstract principle to create something real. Although I was a good student in engineering school, I never felt I had a depth of knowledge until I went to work and was forced to use my theoretical knowledge in a practical way. The result was that I discovered I really knew nothing. I had to go back and relearn from scratch. It was a painful and slow process to relearn much more deeply some of the engineering principles that I thought I knew. But it was worth it, because as a result I could understand and internalize the knowledge and apply it to any number of new situations. My newfound knowledge became a tool in my tool kit. As inventors, we need to have a well-stocked kit of tools that we are familiar with and can use when we need them. Building your tool kit for inventing is a lifelong experience, and it is incumbent upon the inventor to be constantly learning and adding new tools to the kit.

Making It Real

The art of invention involves not only harnessing your creativity and employing it to create novel ideas, but also having the resolve to make those ideas into something real. The process of bringing an idea to fruition can be tortuous and time consuming. It can also lead to tremendous new insight regarding your initial inventive idea. My advice is not to rush through this process. Try to see it as an adjunct to your initial creative flow. There are many jewels to be discovered through the process of going from an idea to an actual working prototype. No matter how brilliant your initial inventive idea is, it will inevitably change and grow as you make it into a working reality. Look at this process as not only giving your idea life, but also as letting it develop and mature as you form it into an actual invention.

The Art and the Science

Is inventing solely an art? In the grand sense, yes—but not exclusively. As you have seen in the course of reading this book, you will need to be both methodical and intellectually rigorous to harness your creative power. Much hard work must come before spontaneous creative thought. The science of the inventive process includes the methods you use to attack a problem and absorb everything there is to know about it as you build your expertise. This can take weeks, months, or even years of work. The science is also how you flesh out and evaluate ideas in a rational and objective manner. It is how you test your invention and draw conclusions from the results.

The art of invention is more subtle. It is developing the sensitivity to harness the power of your unconscious mind and use it effectively. It is making connections that are not readily apparent, but once made, they become obvious in hindsight. The art is moving back and forth between intuition and rational thought—balancing your "feel" and your intellect. The art and the science of invention are joined together. Neither avenue can be successful without the other. Both the beauty and the challenge of inventing is that you need to master both. This is true in almost all creative endeavors and even more so in inventing, where a thorough technical knowledge and deep understanding of the problem you wish to solve is essential to unlocking creative solutions.

THE FUTURE IS YOURS TO CREATE

Inventing is a personal quest. It is totally absorbing and transforms both the intellect and the psyche. It is driven by a consuming desire to create, to realize a vision. Throughout this book, we have examined the many aspects of creating an invention. I started with my personal story, that of a young boy wandering around the aisles of trade shows with his father, going from booth to booth, and examining with wonder all the amazing things that were being displayed. From the perspective of a child, every item, whether a door lock mechanism, a dry ice machine, or an electric toothbrush, was a fascinating object to observe and try to understand. How does it work? What could be done with it? What if . . . ? As adults, most of us would simply give these commonplace objects a quick glance and move on. However, the art of invention requires that we become that child again, that we see the wonder in all things, that we create catalogs in our minds of all that we encounter. We see what is and envision what could be.

The ability to invent is first and foremost a mindset. It is a way of looking at the world differently. You have stepped off the well-trodden path, and each new step requires confidence and belief in yourself that you can do it. And the truth is that you can. The future is yours to create.

Now, go and invent.

ACKNOWLEDGMENTS

As I was writing this book, I was thinking of calling it "Notes to Myself." It would have remained simply that without the encouragement and guidance of many individuals who persuaded me to seek wider horizons. I first must thank my wife, Laura, for reading and rereading the manuscript and telling me how bad my grammar was, but also telling me how much she liked the book. Without Laura, this and many other projects would never have come to fruition. Second, I want to thank Jon Greenberg, who provided both detailed editing and encouragement to seek a wider audience for the book through publication. I want to thank David Sarna for providing numerous contacts and showing me the ropes with regard to publishing. Julianne and Ivan Boden were invaluable in helping me with graphics and photography. The images in this book are a tribute to their creativity and artistic eyes. My close friends David and Michele Morse provided moral support and editing guidance that proved to be invaluable. They were my first test audience once the manuscript was complete. Patent attorneys Arthur Jacob, Gary Walpert, and Gene and Renee Quinn were extremely helpful in reviewing and providing material for my section on patents. I want to thank the intrepid staff at Prometheus Books for all their hard work turning my manuscript into a finished book. In particular, I want to thank editor in chief Steven Mitchell for seeing the merit in this work and my editor for this project, Julia DeGraf, for her critical insight into the text. Although many people have reviewed this book for accuracy, any inadvertent errors in this book are mine and mine alone.

This book was conceived as a tribute to my father, Edward Paley, who, as I have mentioned throughout the book, was my inspiration and teacher in the art of invention. He and my mother, Florence, founded a small company in the basement of our house in 1963 that initially sold products through mail order and eventually went on to become a large multinational corporation. I had the privilege of running this company with my two brothers, Bill and Doug. We served together as the chief executive office of the company and rotated the presidency. This was not only a terrifically successful collaboration but also an inspired learning experience without which this book would not be possible.

Finally, I want to thank my children, Rachel, Shani, Yonah, and Alex, for being encouraging throughout this process, being proud of their father, and keeping the volume down on their video games while I was writing. Thanks, guys.

NOTES

CHAPTER 1

1. "History of the Paper Clip," Early Office Museum, http://www.office museum.com/paper_clips.htm (accessed November 2006).

2. Henry Petroski, *The Evolution of Useful Things* (New York: Vintage, 1992), p. 60.

3. James Adams, *Conceptual Blockbusting* (New York: Norton, 1979), p. 7.

4. Tim O'Reilly, "My Conversation with Jeff Bezos," O'Reilly online, March 2, 2000, http://oreilly.com/pub/a/oreilly/ask_tim/2000/bezos_0300.html (accessed November 2006).

5. Alan C. Kay, "Predicting the Future," *Stanford Engineering* 1, no. 1 (Autumn 1989): 1–6.

CHAPTER 2

1. Edward C. Lathem, *The Poetry of Robert Frost* (New York: Holt Rinehart and Winston, 1969), p. 105. Originally published by Robert Frost in *Mountain Interval* (New York: Henry Holt and Company, 1920).

2. Thomas Stephens, "How a Swiss Invention Hooked the World," Swiss info.ch, January 4, 2007, http://www.swissinfo.ch/eng/search/detail/How%20a%20Swiss%20invention%20hooked%20the%20world.html?siteSect=881&sid=7402384&cKey=1167927120000 (accessed March 2007); "The Invention of Velcro," University of Mary Washington Web site, http://www.umw.edu/hisa/resources/Student%20Projects/Susan%20Deedrick%20—%20Velcro/students.umw.edu/_sdeed5pn/Invention.html (accessed March 2007); Velcro USA Inc., http://www.velcro.com/index.php?page=company (accessed April 2007); Stephen Van Dulken, *Inventing the 20th Century* (New York: New York University Press, 2002), p. 144.

3. Thomas Kuhn, *The Structure of Scientific Revolutions* (Chicago: University of Chicago Press, 1996).

4. Alan C. Kay, "Predicting the Future," *Stanford Engineering* 1, no. 1 (Autumn 1989): 1–6.

5. Tom Monte, *Pritikin: The Man Who Healed America's Heart* (Emmaus, PA: Rodale Press, 1988), pp. 182–83.

6. O. T. Benfey, "August Kekulé and the Birth of the Structural Theory of Organic Chemistry in 1858," *Journal of Chemical Education* 35 (January 1958): 21–23.

7. Kenneth Brown, *Inventors at Work* (Redmond, WA: Microsoft Press, 1988), p. 113.

8. Robert H. McKim, *Experiences in Visual Thinking* (Belmont, CA: Brooks/Cole, 1980), p. 125.

Chapter 3

1. "Dr. Sigmund Freud Dies in Exile at 83," *New York Times*, September 24, 1939.

2. Kenneth Brown, *Inventors at Work* (Redmond, WA: Microsoft Press, 1988), pp. 124–25.

3. Ibid., p. 159.

4. Ibid., pp. 310–11.

5. "Dr. Sigmund Freud Dies in Exile at 83."

6. Carl Jung, *Memories, Dreams, Reflections* (New York: Vintage, 1961), p. 326.

7. Robert H. McKim, *Experiences in Visual Thinking* (Belmont, CA: Brooks/Cole, 1980), p. 38.

8. Michael D. Gershon, *The Second Brain* (New York: HarperCollins, 1998), p. 17.

9. Igor Stravinsky and Robert Craft, *Themes and Conclusions* (Berkeley: University of California Press, 1983), p. 119.

10. Temple Grandin, *Thinking in Pictures, Expanded Edition: My Life with Autism* (New York: Vintage, 2006), pp. 4–5.

CHAPTER 4

1. The initial creative burst of an invention illuminates possibility. To take possibility to reality requires another process called engineering. While invention can be based upon creative leaps or sudden recognition, engineering is a process of planning, setting clear goals, and solving a succession of challenges along the way. It is an arduous process, the hard work behind making a vision come to life. The Patient Care Terminal came to life through the efforts of a group of talented engineers and software developers. It took almost two years from my fortunate lunch date to produce a manufacturable product.

2. Kenneth Brown, *Inventors at Work* (Redmond, WA: Microsoft Press, 1988), p. 224.

3. Tom Monte, *Pritikin: The Man Who Healed America's Heart* (Emmaus, PA: Rodale Press, 1988), pp. 24–25.

4. Brown, *Inventors at Work*, p. 291.

5. Genrich Altshuller, *40 Principles: TRIZ Keys to Technical Innovation* (Worcester, MA: Technical Innovation Center, 1997), p. 9. A list of additional TRIZ publications can be found at http://www.triz.org.

6. Louis Pasteur, lecture given at the University of Lille, December 7, 1854.

7. John Gallawa, "A Brief History of the Microwave Oven," Microtech Web site, http://www.gallawa.com/microtech/history.html (accessed September 2009).

8. James Tobin, *To Conquer the Air: The Wright Brothers and the Great Race for Flight* (New York: Free Press, 2003), p. 70.

9. Judah Ginsberg, "The Discovery of Camptothecin and Taxol," American Chemical Society online, http://acswebcontent.acs.org/landmarks/landmarks/taxol/res.html (accessed October 2009).

10. "Navy Declassifies Details of Pigeon Guidance Project," *Electronic Design*, November 25, 1959, p. 16.

11. D. C. Davies, "Catseyes," BBC H2G2, December 14, 2004, http://www.bbc.co.uk/dna/h2g2/A3320939.

12. Janine Benyus, *Biomimicry: Innovation Inspired by Nature* (New York: William Morrow, 1997), pp. 100–101.

13. "Jaws of Clamworm Are Hardened by Zinc Say UCSB Scientists," University of California, Santa Barbara, August 1, 2003, http://www.ia.ucsb.edu/pa/display.aspx?pkey=1012 (accessed October 2009).

14. Benyus, *Biomimicry*, p. 119.

15. David Stauth and Kaichang Li, "Nature Provides Inspiration for Important New Adhesive," EurekAlert! AAAS, April 11, 2005, http://www.eurekalert.org/pub_releases/2005-04/osu-nii040805.php (accessed May 2007).

16. Joseph Jones, *Robot Programming, A Practical Guide to Behavior-Based Robotics* (New York: McGraw-Hill, 2004), pp. xiii–xiv.

CHAPTER 5

1. Will Schutz, *Profound Simplicity* (San Diego: Learning Concepts, 1982), p. 54.

2. Ibid., p. 55.

3. Michael Hiltzik, *Dealers of Lightning* (New York: HarperCollins, 1999), p. 137.

4. John Maeda, *The Laws of Simplicity* (Cambridge, MA: MIT Press, 2006), p. 20.

5. Stephen Van Dulken, *Inventing the 20th Century* (New York: New York University Press, 2000), p. 178.

6. "Amazon One-Click Shopping," CS201, Stanford University, June 5, 2000, http://www-cs-faculty.stanford.edu/~eroberts/courses/cs181/projects/1999-00/software-patents/ (accessed August 2010).

7. Ibid.

8. Maeda, *The Laws of Simplicity*, pp. 45–46.

9. David Thornburg, *Zero Mass Design*, unpublished manuscript. The manuscript *Zero Mass Design* served as the basis for a course at Stanford University given by Dr. Thornburg in 1983. All references are used with his permission.

10. Ibid., pp. 2, 4.

CHAPTER 6

1. Jack Kilby, "What If He Had Gone on Vacation," interview, Texas Instruments Web site, http://www.ti.com/corp/docs/kilbyctr/vacation.shtml (accessed October 2008).

2. David Harris, "Conix—A New Inhaler for Dry Powders," *Drug Delivery Technology* 7, no. 1 (January 2007): 25–28.

3. Kirsty Barnes, "Dry-Powder Inhaler 'Revolution' Unveiled," in-Pharma Technologist.com, October 24, 2006, http://www.in-pharmatechnologist .com/Packaging/Dry-powder-inhaler-revolution-unveiled (accessed October 2009).

4. "Wikipedia: About," Wikipedia, http://en.wikipedia.org/wiki/ Wikipedia :About (accessed October 2008).

5. Ibid.

6. Mark Yim, Wei-Min Shen, et al., "Modular Self-Reconfigurable Robot Systems," *IEEE Robotics and Automation* 14, no. 1 (March 2007): 43–52.

CHAPTER 7

1. Paul Baran, "On Distributed Communications Networks," *IEEE Transaction on Communications Systems* 12, no. 1 (March 1964): 1–9; Katie Hafner and Mathew Lyon, *Where Wizards Stay Up Late* (New York: Simon & Schuster, 1996), pp. 56–59.

2. Matt Ford, "A Homeowner's Dream: A House That Fixes Itself," *Ars Technica* (April 3, 2007), http://arstechnica.com/journals/science.ars/2007/04/ 03/ahome-owners-dream-a-house-that-fixes-itself (accessed November 2008).

3. Sally Adee, "Self-Healing Hulls," *IEEE Spectrum* 45, no. 11 (November 2008): 16.

4. Peter Weiss, "Unstoppable Bot: Armed with Self-Scrutiny, a Mangled Robot Moves On," *Science News* 170, no. 21 (November 18, 2006): 324.

5. Joseph Jones, *Robot Programming, A Practical Guide to Behavior-Based Robotics* (New York: McGraw-Hill, 2004), pp. 83–87.

CHAPTER 9

1. John Dessauer, *My Years with Xerox: The Billions Nobody Wanted* (New York: Manor Books, 1971).

2. "Robert Kearns, Inventor of Intermittent Windshield Wipers and Battled Car Companies, Dies at 77," Auto Channel Web site, February 25, 2005, http://www.theautochannel.com/news/2005/02/25/005398.html (accessed January 2008).

3. All dollar amounts in this section are given as of 2009. The amounts are very approximate estimates as costs range widely depending on the patent attorney, the specifics of the invention, and the complexity of prosecution.

4. Gene Quinn, "The Cost of Obtaining a Patent," IPWatchdog.com, http://www.ipwatchdog.com/patent/patent-cost (accessed January 2008).

Chapter 10

1. Walter Isaacson, *Einstein: His Life and Universe* (New York: Simon & Schuster, 2007), p. 113.

2. Dilip Ramchandani, "The Librium Story," http://www.benzo.org.uk/librium.htm (accessed September 2009).

3. Jonathan Cohen, in discussion with the author, September 2009. Dr. Jonathan Cohen was a personal friend and colleague of Dr. Leo Sternbach. Dr. Cohen, a research psychiatrist, directed many of the clinical trials of Valium.

BIBLIOGRAPHY

Adams, James L. *Conceptual Blockbusting: A Guide to Better Ideas.* New York: Norton, 1979.

Adee, Sally. "Self-Healing Hulls." *IEEE Spectrum* 45, no. 11 (November 2008): 16.

Altshuller, Genrich. *40 Principles: TRIZ Keys to Technical Innovation.* Worcester, MA: Technical Innovation Center, 2002.

———. *And Suddenly the Inventor Appeared.* Worcester, MA: Technical Innovation Center, 2004.

"Amazon One-Click Shopping." CS201, Stanford University, June 5, 2000. http://www-cs-faculty.stanford.edu/~eroberts/courses/cs181/projects/1999-00/software-patents/ (accessed August 2010).

Baran, Paul. "On Distributed Communications Networks." *IEEE Transaction on Communications Systems* 12, no. 1 (March 1964): 1–9.

Barnes, Kirsty. "Dry-Powder Inhaler 'Revolution' Unveiled." in-Pharma Technologist.com. http://www.in-pharmatechnologist.com/Packaging/Dry-powder-inhaler-revolution-unveiled (accessed October 2009).

Bellis, Mary. "The Invention of VELCRO®—George de Mestral." About.com: Inventors. http:/inventors.about.com/library/weekly/aa091297.htm (accessed December 2006).

Benfey, O. T. "August Kekulé and the Birth of the Structural Theory of Organic Chemistry in 1858." *Journal of Chemical Education* 35 (January 1958): 21–23.

Benyus, Janine M. *Biomimicry: Innovation Inspired by Nature.* New York: William Morrow, 1997.

Bongard, Josh, Victor Zykov, and Hod Lipson. "Evolutionary Robotics." Spatial Robots.com, August 4, 2009. http://www.spatialrobots.com/2009/08/evolutionary-robotics-by-josh-bongard-victor-zykov-and-hod-lipson/ (accessed October 2009).

Brown, Kenneth A. *Inventors at Work.* Redmond, WA: Tempus Books, 1988.

"Chicle Gum a Truly American Chew." NPR.org, June 12, 2009. http://www.npr.org/templates/story/story.php?storyId=106521646 (accessed October 2009).

"The Chip That Jack Built." Texas Instruments Web site. http://www.ti.com/corp/docs/kilbyctr/jackbuilt.shtml (accessed October 2008).

Csikszentmihalyi, Mihaly. *Flow: The Psychology of Optimal Experience.* New York: Harper & Row, 1990.

Davies, D. C. "Catseyes." BBC H2G2, December 14, 2004. http://www.bbc.co.uk/dna/h2g2/A3320939 (accessed October 2009).

Dessauer, John. *My Years with Xerox: The Billions Nobody Wanted.* New York: Manor Books, 1971.

Eberhart, Mark. *Why Things Break.* New York: Harmony Books, 2003.

"Editors' Choice: Best Products of 1990." *Semiconductor International* (December 1990): 20–23.

Evans, Harold. *They Made America: From the Steam Engine to the Search Engine.* New York: Back Bay Books/Little, Brown, 2004.

"Evolutionary Robotics." Cornell Computational Synthesis Laboratory Web site. http://ccsl.mae.cornell.edu/evolutionary_robotics.

Ford, Matt. "A Homeowner's Dream: A House That Fixes Itself." *Ars Technica* (April 3, 2007). http://arstechnica.com/journals/science.ars/2007/04/03/ahome-owners-dream-a-house-that-fixes-itself (accessed November 2008).

"Four Cent Inhaler Offers Real Alternative for Pandemic Vaccination." Cambridge Consultants Web site, October 24, 2006. http://www.cambridgeconsultants.com/news_pr177.html (accessed October 2009).

Gallawa, John. "A Brief History of the Microwave Oven." Microtech Web site. http://www.gallawa.com/microtech/history.html (accessed September 2009).

Gershon, Michael D. *The Second Brain.* New York: HarperCollins, 1998.

Ginsberg, Judah. "The Discovery of Camptothecin and Taxol." American Chemical Society online. http://acswebcontent.acs.org/landmarks/landmarks/taxol/res.html (accessed October 2009).

Grandin, Temple. *Thinking in Pictures, Expanded Edition: My Life with Autism.* New York: Vintage, 2006.

Hafner, Katie, and Matthew Lyon. *Where Wizards Stay Up Late: The Origins of the Internet.* New York: Simon & Schuster, 1996.

Harris, David. "Conix—A New Inhaler for Dry Powders." *Drug Delivery Technology* 7, no. 1 (January 2007): 25–28.

Hiltzik, Michael. *Dealers of Lightning: Xerox PARC and the Dawn of the Computer Age.* New York: HarperCollins, 1999.

"The History of Chicle Chewing Gum." Gumballs.com. http://www.gumballs.com/history-of-chicle-chewing-gum.html (accessed October 2009).

"The History of the Integrated Circuit." Nobelprize.org. http://nobelprize.org/educational_games/physics/integrated_circuit/history/index.html (accessed October 2009).

"History of the Paper Clip." Early Office Museum. http://www.officemuseum.com/paper_clips.htm (accessed November 2006).

The Industrial Erector Set. 80/20 Inc. http://www.8020.net/Default.asp (accessed August 2007).

InnoCentive. http://www2.innocentive.com (accessed November 2009).

"The Invention of Velcro." University of Mary Washington Web site. http://www.umw.edu/hisa/resources/Student%20Projects/Susan%20Deedrick%20—%20Velcro/students.umw.edu/_sdeed5pn/Invention.html (accessed March 2007).

Isaacson, Walter. *Einstein: His Life and Universe.* New York: Simon & Schuster, 2007.

"Jaws of Clamworm Are Hardened by Zinc Say UCSB Scientists." University of California Santa Barbara, August 1, 2003. http://www.ia.ucsb.edu/pa/display.aspx?pkey=1012 (accessed October 2009).

Jones, Joseph L. *Robot Programming: A Practical Guide to Behavior-Based Robotics.* New York: McGraw-Hill, 2004.

Jung, Carl. *Memories, Dreams, Reflections.* New York: Vintage, 1961.

Kay, Alan C. "Predicting the Future." *Stanford Engineering* 1, no. 1 (Autumn 1989): 1–6.

Koelbing, H. M. "Chance Favors the Prepared Mind Only." *Inflammation Research* 1, no. 1 (July 1969): 53–56.

Kuhn, Thomas. *The Structure of Scientific Revolutions.* Chicago: University of Chicago Press, 1996.

Lidwell, William, Kritina Holden, and Jill Butler. *Universal Principles of Design.* Gloucester, MA: Rockport Publishers, 2003.

Lindsay, Andy. *Robotics with the Boe-Bot.* Rocklin, CA: Parallax, 2004.

Maeda, John. *The Laws of Simplicity.* Cambridge, MA: MIT Press, 2006.

"Masks and Reticles." SiliconFarEast.com. http://www.siliconfareast.com/masksreticles.htm (accessed March 2007).

McKim, Robert H. *Experiences in Visual Thinking.* Belmont, CA: Brooks/Cole, 1980.

Moggridge, Bill. *Designing Interactions.* Cambridge, MA: MIT Press, 2007.

Monte, Tom. *Pritikin: The Man Who Healed America's Heart.* Emmaus, PA: Rodale Press, 1988.

Murata, Satoshi, and Haruhisa Kurokawa. "Self-Reconfigurable Robots." *IEEE Robotics and Automation* 14, no. 1 (March 2007): 71–78.

National Inventors Hall of Fame Foundation. *Invent Now.* http://www.invent.org/about_invent_now/4_0_0_about.asp (accessed August 2008).

"Navy Declassifies Details of Pigeon Guidance Project." *Electronic Design* (November 25, 1959): 16.

"1958: Invention of the Integrated Circuit." PBS.org. http://www.pbs.org/transistor/background1/events/icinv.html (accessed October 2008).

O'Reilly, Tim. "My Conversation with Jeff Bezos." O'Reilly online, March 2, 2000. http://oreilly.com/pub/a/oreilly/ask_tim/2000/bezos_0300.html (accessed November 2006).

Papanek, Victor. *Design for the Real World.* New York: Bantam Books, 1973.

"Percy Shaw." Design Museum online. http://designmuseum.org/design/percy-shaw (accessed October 2009).

Petroski, Henry. *The Evolution of Useful Things.* New York: Vintage, 1992.

Quinn, Gene. "The Cost of Obtaining a Patent." IPWatchdog.com. http://www.ipwatchdog.com/patent/patent-cost (accessed January 2008).

Ramchandani, Dilip. "The Librium Story." http://www.benzo.org.uk/librium.htm (accessed September 2009).

"Robert Kearns, Inventor of Intermittent Windshield Wipers and Battled Car Companies, Dies at 77." Auto Channel Web site, February 25, 2005. http://www.theautochannel.com/news/2005/02/25/005398.html (accessed January 2008).

Schutz, Will. *Profound Simplicity.* San Diego, CA: Learning Concepts, 1982.

"Self-Modeling Robotics: Movies and Pictures." Cornell Computational Synthesis Laboratory Web site. http://ccsl.mae.cornell.edu/research/self-models/morepictures.htm.

"Self-Replication: More Movies and Pictures." Cornell Computational Synthesis Laboratory Web site.http://ccsl.mae.cornell.edu/research/selfrep/morepictures.htm.

Sherer, Kyle. "UK Researchers Developing Self-Repairing Aircraft." Gizmag online, May 27, 2008. http://www.gizmag.com/self-repairing-plane/9385/ (accessed January 2009).

Slowiczek, Fran, and Pamela M. Peters. "Discovery, Chance and the Scientific Method." Access Excellence Web site. http://www.accessexcellence.org/AE/AEC/CC/chance.php (accessed October 2009).

"Springs." How Products Are Made. Madehow.com. http://www.madehow.com/Volume-6/Springs.html (accessed August 2007).

Stauth, David, and Kaichang Li. "Nature Provides Inspiration for Important New Adhesive." EurekAlert! AAAS, April 11, 2005. http://www.eurekalert.org/pub_releases/2005-04/osu-nii040805.php (accessed May 2007).

Stephens, Thomas. "How a Swiss Invention Hooked the World." Swissinfo.ch, January 4, 2007. http://www.swissinfo.ch/eng/search/detail/How%20a%20Swiss%20invention%20hooked%20the%20world.html?siteSect=881&sid=7402384&cKey=1167927120000 (accessed March 2007).

Strauss, Steven. *The Big Idea.* Chicago: Dearborn Trade Publishing, 2002.

Taleb, Nassim N. *The Black Swan.* New York: Random House, 2007.

Technical Innovation Center. *Systematic Innovation with TRIZ.* http://www.triz.org/ (accessed February 2007).

Thornburg, David. *Zero Mass Design.* Manuscript and lecture. Stanford University, 1983.

Tobin, James. *To Conquer the Air: The Wright Brothers and the Great Race for Flight.* New York: Free Press, 2003.

Van Dulken, Stephen. *Inventing the 20th Century.* New York: New York University Press, 2000.

Velcro USA Inc. http://www.velcro.com/index.php?page=company (accessed April 2007).

Waldrop, Mitchell. *The Dream Machine.* New York: Viking, 2001.

Weiss, Peter. "Unstoppable Bot: Armed with Self-Scrutiny, a Mangled Robot Moves On." *Science News* 170, no. 21 (November 18, 2006): 324.

"What If He Had Gone on Vacation." Texas Instruments Web site. http://www.ti.com/corp/docs/kilbyctr/vacation.shtml (accessed October 2008).

"Wikipedia: About." Wikipedia. http://en.wikipedia.org/wiki/Wikipedia:About (accessed October 2008).

Yim, Mark, Wei-Min Shen, et al. "Modular Self-Reconfigurable Robot Systems." *IEEE Robotics and Automation* 14, no. 1 (March 2007): 43–52.

INDEX